T0302167

Leadership Skills for Maintenance Supervisors and Managers

Leadership Skills for Maintenance Supervisors and Managers

Joel D. Levitt

CRC Press is an imprint of the
Taylor & Francis Group, an **informa** business

First edition published 2020
by CRC Press
6000 Broken Sound Parkway NW, Suite 300, Boca Raton, FL 33487-2742

and by CRC Press
2 Park Square, Milton Park, Abingdon, Oxon, OX14 4RN

CRC Press is an imprint of Taylor & Francis Group, LLC

Library of Congress Cataloging-in-Publication Data

Names: Levitt, Joel, 1952- author.
Title: Leadership skills for maintenance supervisors and managers / Joel D. Levitt.
Description: First edition. | Boca Raton, FL : CRC Press, 2021. | Includes bibliographical references and index.
Identifiers: LCCN 2020036372 (print) | LCCN 2020036373 (ebook) | ISBN 9780367481759 (hardback) | ISBN 9781003097952 (ebook)
Subjects: LCSH: Industrial management. | Supervisors--Training of. | Employee maintenance. | Organizational effectiveness.
Classification: LCC HD31.2 .L48 2021 (print) | LCC HD31.2 (ebook) | DDC 658.2/02--dc23
LC record available at https://lccn.loc.gov/2020036372
LC ebook record available at https://lccn.loc.gov/2020036373

ISBN: 978-0-367-48175-9 (hbk)
ISBN: 978-1-003-09795-2 (ebk)

Typeset in Times
by SPi Global, India

Visit the CRC Press website for PowerPoint slides: https://www.routledge.com/Leadership-Skills-for-Maintenance-Supervisors-and-Managers/Levitt/p/book/9780367481759

Our Mission

We are committed to helping organizations worldwide treat all people and the environment with respect.

Contents

SECTION IV The Future of Managing Maintenance

Preface

We tend to train maintenance supervisors by throwing them into the deep end of the pool and seeing if they can survive. It has worked for generations reasonably well. It is also wasteful, disruptive, and unnecessarily painful. There is a significant risk with the sink or swim method. In some cases, we take a great tradesperson and turn them into a mediocre supervisor and take a relatively happy person and turn them unhappy.

What this book seeks to do is twofold. The first is to provide a comprehensive training manual for new supervisors and a refresher for seasoned supervisors. The second is to give potential supervisors a taste of the field with a realistic view of the challenges and rewards.

It was evident in 1990 when I started training maintenance managers that training was a low priority for their supervisors. I asked my maintenance manager students what was missing from their training and preparation for their first leadership role. They were eager to share their experience and their lack of preparation. Like a lot of people who went through a tough apprenticeship (I'm thinking about Doctors), they are rightfully proud of having survived.

There seems to be no recognition of the unique roles of maintenance in general and maintenance supervision in particular. One day the supervisor is leading the crew to safely repair a dangerous breakdown and the next, dealing with a sensitive personnel issue. Maintenance folks are usually well prepared for the fixing part of the job and utterly unprepared for the human element of their job.

That a decent supervisor must be technically competent is afore given conclusion. That they also have to be therapists, security officers, teachers, judges, friends, cheerleaders, executioners are usually a surprise. But the supervisor must supply what is needed at the moment to get the job done.

This book has four major sections:

Psychology of Supervision and Subordinates: Once a candidate becomes a supervisor, they trigger psychology of authority, relationships, personalities that are wired into people's minds and sometimes take the new supervisor by surprise. Things changed when they became a supervisor and it is not clear why!

Maintenance Management for supervisors: Another aspect of maintenance supervision is that the advances, new approaches, insights, and technology of maintenance management are mainly opaque to them. They are generally not the first ones to be chosen for conferences, trade shows, and management training. As a result, they are entering a "thought" world that uses different languages, has unique models, and has new incomprehensible tools.

Supervisor's Toolbox: Everyone wants useful tools and tricks to help them do their job. Topics covered range from supervising friends (and enemies) to team building, from the quality of work, and avoiding mistakes to time management. The section contains tools, ideas that will make the supervisor's life easier.

Making technology, your friend: A tidal wave of new technology is rocking the maintenance world. This section gives the supervisor a capsule introduction to the most important of the techniques, technologies, software, and hardware.

Additional eResources that accompany this book are are also on the Routledge website:www.routledge.com/9780367481759. If you are a new supervisor, congratulations. You will have an impact on your people well beyond the job. People will grow and develop just because of your attention and intention.

Joel Levitt, 2020

Author

Joel D. Levitt CMRP, CRL, CPMM, has been a Certified Prosci Change Management practitioner since 1980. Mr. Levitt has also been the president of Springfield Resources, maintenance consultants, and trainers in a wide variety of industries, including pharmaceuticals, oil, airports, hospitals, high tech manufacturing, school systems, and government. He is a leading maintenance trainer throughout the US, Canada, Australia, Europe, and Asia.

He has trained over 17,000 maintenance professionals from 39 countries in more than 600 classes and workshops. Surveys of his students show that 98% rated the training very good or excellent. Mr. Levitt was the director of International Projects at Life Cycle Engineering and director of Reliability Projects at Reliabilityweb and a trainer of the CRL. He was a senior consultant at Computer Cost Control Corp. There he designed computerized maintenance management systems for organizations including FedEx, United Airlines, JFK Airport, and BFI. He designed, installed, and serviced a complete automation system with rack controls, accounting, and inventory controls for BP North America's 30,000-barrel/day-oil terminal.

Mr. Levitt has written 14 books on maintenance management and contributed chapters to two other books. He has written over 200 articles for trade publications on maintenance topics. His teaching style is humorous and understandable straight talking. Students should expect to be introduced to a great deal of material, real-life examples, successful practices from around the world, and have a good time as well. Typical clients include: GM, BHP, Volvo, USX, Mercedes Benz, Valero Oil X, ARAMCO, Lawrence Livermore National Laboratory, Esso Canada, Chevron-Saudi Arabia, Arawak Cement-Barbados, BP, Atlantic LNG, Iron Ore of Canada X, LA Water, Wyeth, Saint Martin International Airport, Dubai Water, US Army Corps of Engineers (Dams on the Columbia, Red River), Portland International Airport, CUF (Ammonia Plant), Syncrude, Algoma X, GE Plastics (Now SABIC), Abbotts Labs, Merck, Prince George Pulp and paper, City College of NY, Univ. of Alabama, Ft Meade (US Army), SAP, Jefferson University Hospital, US Defense Logistics, Philip Morris, Sony, CSX Railroad, Harley Davidson Motorcycle, US Navy, and 3000 more.

Introduction

GENESIS OF THIS BOOK

I started teaching maintenance managers over 32 years ago for Clemson University. About 30 years ago, I asked my classes of maintenance managers what was missing in training from when they transitioned from worker to supervisor. I also asked them what wisdom they would like to impart to their younger selves.

At the time, I designed a course to cover the areas the managers mentioned. Management Skills for Maintenance Supervisors and its sequel Management Strategies for Maintenance Supervisors were the result. Those two classes were given in various forms for over 29 years at the University of Alabama (and elsewhere) to 1000s of supervisors and managers.

Comments in 1990 Answers to the question: What are the attributes of a great maintenance supervisor?

We felt that there already was significant knowledge about leadership within the maintenance community. We wanted to tap into this rich tradition. The answers fell into three general categories:

People	Management	Technical
Good listener	Organized	Dedicated to quality
Compassionate	Ability to make decisions	Knows equipment
Can motivate others	Good delegator	Knows the job
Fair and consistent	Meets goals of business unit	Knows safety
Respected	Re-analyzes progress to goal	Can analyze problems
Honest	Knows what is and isn't important	Can evaluate the skill level
Innovator	Provides good service to customer Understands product	
Open-minded	Loyal to organization	
Effective communicator	Oriented toward results	
A coach, not a dictator	Good planner	
Good negotiator	High productivity	
Positive outlook	Follows up to see the job is done	
Flexible	Understands the importance of scheduling	
Treats people as equals	Assign and keeps priorities	
Can read people	Understands and uses budgets	
Adaptable to change		
Common sense	Provides a conduit for downward communication	
Willing to learn	Provides intelligence to upper management	

- Not afraid to make mistakes
- Will take control if necessary
- Effective trainer, communicator
- Can work with different types of people

- Gives recognition for jobs well done
- Can deal with difficult people issues
- Praises in public, disciplines in private
- Has a cool head; can handle the pressure

Notice that most of the comments from these managers centered on people issues and management issues. Many people have an idea about maintenance leadership and management to exhibit some of the attributes they discussed at least some of the time. In the next question, the interviewees passed along some of the wisdom learned from their years on the job.

We asked people to pass along gems of wisdom to new supervisors, and these were their favorites:

- Be a good listener.
- Learn to bend, but don't abdicate.
- Remember, you are in charge; act that way.
- Strive to be respected, not necessarily liked.
- Always be available to your people.
- Cultivate your patience. Know which things can be put off or ignored and which can't.
- Give clear indications of what a good job is, give praise when appropriate
- Don't be afraid to acknowledge that you don't know.
- Quality is an attitude.
- There is a fine line between getting involved and getting in the way.
- Good supervisors surround themselves with good people and are not afraid of training replacements.
- Keep a positive attitude, keep company interests at heart.
- Set goals every day, review before leaving. Plan your days.
- Listen more, talk less. Be able to hear feedback you don't like.
- Solicit the views of the workers for improvements and problem areas.
- Use positive, one-on-one techniques with workers.
- Follow your work plan.
- Keep your ass covered.
- Pay attention to who your people are
- Never stop learning.
- Treat people consistently, fairly and firmly.
- Keep your eyes open; don't just look but **see**.
- Make time to analyze your problem areas, compile facts before deciding.
- Those tough, humbling experiences are valuable, treasure them.

Some of specific comments 2019 were as follows:

In 2019, while working on this book, I conducted a similar survey on LinkedIn. Not surprisingly, the answers were very similar. Note that few people asked for more training in technology, social media, and only one mentioned computerization. I would like to thank my LinkedIn community for their insights and help.

According to Terry Drinkard, Senior Structural Engineering supervisor; supervision is not about profits or products or schedules, it's about having integrity and demonstrating that every single day.

Bermla (Kata) Yawing, Mechanical Supervisor at Newcrest Mining Limited - Lihir Mine Maintenance advice is fundamental, teach time management, and he would actively go for these three things to be a success with my team.

1. Safety: Do the job safely.
2. Quality: To the required standard.
3. Efficiency: In the most efficient manner possible.

Donald Metcalf Master Electrician/Controls Technician

1. Anything on time management would have been excellent training. Any kind of teaching training. You are teaching to adults who are a different animal than younger students.
2. The most useful resource, in my opinion, is mentorship. Nothing can replace having someone show you how to do and then stand by while you do until you are competent.
3. How to deal with people. Everyone is different and requires different strategies.
4. Also, the phrase "I can do anything, just not everything!" There is only one you! Delegate!
5. And treat everyone as you would like to be treated, with respect.
   Everyone spoke in the wheel is critical! From the janitor to the CEO.

Dave Reiber

1. I would have liked some insight into specific tools required to be proficient. Things like necessary data needed for accurate decision making, from CMMS, Condition monitoring, technicians' input, budgeting information, (like Capex to Opex and Margins). Also, Asset Criticality Analysis associated with Spare parts availability and lead times,
2. Mentorship is first and foremost. There is no better teacher than someone who has been there, done that.
3. I would have liked someone to let me know that the toughest things that I would encounter would always be people skills. No matter what we deal with on an average day, the relationships and the strengths or weaknesses of them, determine the results. When you build strong teams, you can expect higher innovation and creativity. You can expect people that will go the extra distance when required, and you can expect long term alliances that will support your journey.

Bill Sugden, I would favor starting with a fundamental understanding of:

1. Consequence/s of asset failure as a basis for prioritizing effort.
2. Failure modes that lead to asset failure and how to detect them.
3. Planning, scheduling & coordination.

4. Leadership and team-building.
5. Root cause analysis to drive improvement. Once we've covered that, I'd think about the next layer.

Duane Ayres, Manager of Water Treatment Operations, at the City of Brantford, advises new supervisors.

1. Knowing the total scope of responsibilities you will have as a supervisor. There is a lot to know, and a lot needed to be in that position.
2. Knowing more about the financial responsibilities you have been a maintenance supervisor would have helped.
3. Class training for Supervisors, leadership training, and my mentorship. My mentor was very detailed, which was great.
4. The best piece of advice ever. GET TO KNOW YOUR PEOPLE.

Tommy Archer, CMRP, CMM, Engineering Manager at Coca-Cola Bottling Company UNITED, Inc. advice ranged from training to realizing the people already there can be great resources.

1. Training on the "Transition from Tools to Technical Leadership" would have changed mine and many others' careers. Companies frequently choose the most technical people to supervise but rarely help them make the transition.
2. My most useful resources were former supervisors and managers. I always tried to glean the right practices and desperately avoid the bad from past leaders. Most of my business/maintenance technical skills came from books and, more recently, podcasts/white papers.

   A mentor would have been great!
3. Wisdom? First, recognize that your people are more than assets to work with, know what motivates them, and help them develop and meet their goals. Second, supervise as if you own the business, expect the best from people, and hold them accountable. Accountability can be positive as well as harmful. Last, be courageous, never be afraid to challenge your people to greatness, and never be afraid to defend them when needed.

Duó Jose Domingos, Gerente de Manutenção (Maintenance Manager), from São Paulo, Brazil contributes,

1. Before starting as Supervisor, receive guidance for at least six months from a professional with experience in the area.
2. The resource I recommend is coaching because it can lead you to reflect, reach conclusions, define actions and, mainly, act towards your goals, goals, and desires.
3. Use empathy. Put yourself in the other's shoes. Learn to listen. Have humility and respect for people and recognize their skills as well as value their opinions. Be a server leader.

Gary Chamberlain, Technical Support at Flightstar Aircraft Services, LLC

1. Authority and time management: I found management dumps everything down on Supervisors and Leads without expressively stating which tasks are directly the Supervisor's responsibilities and which can be delegated.
2. The most useful resource to any supervisor is the team the position oversees. Mentorship would help immensely in breaking in a new supervisor with no previous experience with the crew. Second is strong managers keeping administrative duties from falling further down.
3. Don't be afraid of letting the system fail; if you continually work around systemic problems, those responsible for the structure or process will not become aware of its failures, and they won't be corrected.

To accommodate everyone's ideas in a logical format, I divided this book into four sections:

- The psychology of supervision and subordinates,
- Maintenance management for supervisors,
- The supervisor's toolbox, and
- The future of managing maintenance.

HOW TO USE THIS BOOK?

The sections are designed to gather together information in similar areas and should aid in finding discussion about areas of interest.

Also, I divided the craft of supervision into three major sections: People Skills (psychology section), Management Skills (Maintenance management section), and Technical Skills (Supervisor's toolbox section).

By all means, dip into the book using the index and complete table of contents and read about situations you are facing. Read the whole book to learn about potential issues and how to take steps to avoid the problem.

I designed this book to give supervisors additional tools to improve productivity. Supervisors' productivity is **highly leveraged** because if your productivity improves, your entire crew improves.

The job maintenance supervisor is to get the maintenance work completed on time, on budget, and safely! We can be popular. We can be great communicators. We can be inspired, trainers. We can be great repair people. These things alone do not make successful maintenance supervisors. We are not successful maintenance supervisors unless we can keep the machines running, the fleet rolling, and the occupants warm in the winter.

Judge this work by its potential effect on your bottom line and how it can impact your ability to get the job done.

The day you became a supervisor, you became aware of a secret that all leaders experience. The secret is, "Being a supervisor doesn't make you a supervisor." You must develop your skills at supervision the same way you learn any craft.

You make mistakes, but you can also learn from your mistakes. You can focus attention on the things around you, alert for opportunities to learn. You also may have formal education, attended seminars, read books about maintenance supervision. These all contribute to building an effective supervisor.

This book will focus on methods of dealing with the everyday situations of maintenance supervision. I borrowed these methods from many fields, including engineering, psychology, sociology, mathematics, and organizational development.

LEARNING TO LEARN

There are a few simple techniques to learn more efficiently.

If you are reading a book, watching a video, or listening to a podcast, TED talk or another stream of content that you consider useful consider:

1. Write notes with a format that identifies where in the work you are (page number, minutes into a video), what is the specific idea that caught your attention, and leave a blank for comment or action item. You might have to stop from time to time to do this.

   As the book, video, or seminar progresses, list items that you will change in yourself or your organization or issues which you will investigate toward improvements.

   See the Action Master List (AML) in the appendix. Write a note to yourself at the beginning of each section or chapter to remember this AML.
2. Write a summary of the program and discuss it with others. If you can generate the learning so that others can understand it, then you know the material.
3. Do something different in your organization because of what you learned. Time is of the essence.

See ACTION Master List in the Appendix.

Section I

Psychology of Supervision and of Subordinates

1 What Is Supervision?

It is useful to give some thought as to your role. Ask your manager what he or she expects from you. You also might ask your direct reports what they expect from you.

LEADERSHIP, MANAGEMENT, AND SUPERVISION

In the smallest maintenance departments, you can successfully operate without a supervisor. At coffee in the morning, everyone can informally get on the same page and ferret out the priority of the work on the clipboard (or in-box, pile, or whatever).

The issue is when the department is big enough that there are too many actions at different times in different places for anyone to keep track of and avoid collisions of resources.

To have successful maintenance, you must have a realistic view of supervision and provide the supervisor with the tools, authority, and support they need. Finally, to have successful maintenance, you must have respect for what the supervisor contributes. We will examine the roles and responsibilities for successful supervision and if our companies sabotage success.

In the maintenance field, the supervisor is the critical player in running the team. At a basic level, the supervisor is accountable for the quality, safety, security, and productivity of the maintenance effort. If you read articles about modern supervision, you will see quite a bit of discussion on the supervisor as a leader and the supervisor as a manager. If you read between the lines, you understand that supervisors should be great leaders, efficient managers, and focus on getting the work done.

This expectation is nothing new. The supervisor will play multiple roles for the good of the company and the good of their group. So are supervisors' leaders, managers, or just supervisors? What exactly do we hire supervisors to do? I think there is some confusion about what we want from supervisors.

Maybe there is no confusion at all; we want everything. (Perhaps we are seeking a miracle worker or magician to make up for the lack of our companies' leadership and management!) Before we can discuss this, let's define the three terms.

The leader is the person who leads or commands a group, organization, or country. "Leadership is about mapping out where you need to go to "win" as a team or an organization, and it is dynamic, exciting, and inspiring." (www.mindtools.com)

A manager is a person responsible for controlling or administering all or part of a company or similar organization. "Management is the organization and coordination of the activities of a business to achieve defined objectives." (http://www.businessdictionary.com)

Supervision is the act or instance of directing, managing, or oversight, especially a critical watching and directing (as of activities or a course of action). (www.merriam-webster.com)

In the maintenance business, supervision is overseeing the repair, PM, and small project work.

We ask our supervisors for leadership and to be leaders. If you are an American, leadership is like apple pie or the Fourth of July. It is patriotic; leadership is red-blooded. If you are not American, leadership is the same, translated into your country's symbols and memes. Much of our impression of leadership is from war movies and TV shows. We ask for leadership to take the next mountain, the next bridge. In that kind of leadership, it is entirely OK to sacrifice the leader and their troops to take that hill or bridge; or is it? That is an idealized version of real leadership.

The maintenance leader, the supervisor, is a quieter quality.
In maintenance, the "leader" calls forth the best work from their team members. The leader is "there" for his or her people. The leader looks out for their people, tries to protect them from the bad decisions of the managers, gets them training, recognition, and takes the heat when there are mistakes. They also protect the workers from their tendencies to take shortcuts, be unsafe, or compromise when it comes to environmental or health issues.

The company supports or undermines the supervisor's leadership by allocating adequate funds for training, tools, support systems, and buying the right equipment in the first place.

In the same way, the idea of management comes from images of a relentless cost-cutter, efficiency expert, or strictly a "numbers" person. The idealized manager has no time for soft skills, soft people, or for anything that doesn't directly impact the value stream.

The maintenance supervisor is a quieter kind of manager.
This kind of manager starts early to ensure, as far as possible, the mechanics have everything they need to do their jobs. They make certain other groups that maintenance depends on are ready so that the task can proceed smoothly.

They chase after their team members to make sure all work has work orders and records all hours, all parts, and all other elements of the job.

The company supports or undermines the supervisor's management by allocating adequate funds for CMMS and proper planning and scheduling. They support his or her management by insisting that everyone adheres to the schedule (including the issuance of permits and cleaning equipment before the work is to start). They agree to the importance of the PM/PdM effort and the adherence to that schedule as well. The company also provides adequate support in the form of parts, adequately staffed warehousing, reliability, and maintenance engineering.

Performance

The managers of maintenance must be effective leaders to improve performance. Within the maintenance community, there are excellent, though frequently unheralded, examples of leadership. In the words of A.S. Migs Damiani (Jan '96, *Facilities*

Magazine), a leader accepts the challenge for change. He or she thinks positively and big, knows the business of their business, becomes their bosses' teacher, invests in themselves, focuses on training, increases their visibility, involves their people, has financial know-how, values diversity, and looks for the gold in others. This list is an excellent one to help get our hands around the concept of leadership.

Maintenance departments have legends about heroes, demons, and devils. These stories from the past frequently reflect the values of the institution. Some myths concern heroic deeds of breakdown repair, saving lives, saving products, or machinery. Others involve grand thefts, hateful chiefs, evil purchasing agents, drinking and partying, high stakes card games, or other illegal activities. These legends reveal insight into the underlining patterns governing the department and, possibly, the whole organization.

Maintenance leadership transforms the legends from great repairers to great reengineers, from stories about heroic work through the night to idyllic "it never breaks down while we take care of it" people and stories. When we shift the rewards, stories, and legends, we begin to create the possibility of running a maintenance organization without repair. Today, unlike in the past, authentic maintenance leadership does not focus on improving the efficiency of the arts of repair, but in looking for other ways to preserve the asset and, more importantly, maintain the capability that the asset provides.

Other Traits

In traditional environments, leadership was less critical than the supervisor's other capabilities. In the 1981 edition of *How to Manage Maintenance* by AMA, leadership was the first attribute mentioned, but it was never defined. Their list of considerations for selecting people to be foreman (the gender-neutral word supervisor has superseded the phrase) includes:

1. Who is a leader?
2. Who has demonstrated skills in planning?
3. Who understands the work order and priorities system?
4. Who is an excellent communicator?
5. Who can sketch?
6. Who is respected by peers and superiors?
7. Who is good at figures?
8. Who is an innovator?

This list is still valid today. But many of the tasks mentioned above have been transferred to the trade's people themselves. Thus, leadership is more critical than ever before.

One comment made in the same text was that the supervisor's position is an excellent training ground for young engineers. This choice is a common management misconception, which would be comical if it weren't so frequently disastrous. Put a young and inexperienced person in charge of a workgroup of experienced trade's people? They will have him or her for lunch—daily until the engineer proves him/

herself. Yet it is true that engineers need real experience in maintenance before they are let loose as designers of plants and equipment.

One of the gurus of management, Stephen Covey, has a story in his seminar on *The Seven Habits of Successful People* that demonstrates the difference between leadership and management. A group of people is trying to hack a road through the jungle with machetes. To support them, the managers developed machete sharpeners, apprentice machete wielders, and other systems that made the work smooth and efficient. One of the people climbed a tree and yelled down that they were going the wrong way. The management on the ground told him to shut up because they were making such good progress. In Covey's model, leadership knows where to take the maintenance workgroup and, more importantly, where to take the maintenance function.

Advice from Experts

Scott King, the maintenance supervisor at Forest Home Inc., is quoted in *The Maintenance Supervisor's Standard Manual* as having four specific areas of focus for improved leadership:

1. Spend time with employees. He spends at least 30 minutes a week, one-on-one with each subordinate, and believes that this is the most valuable time he spends. When employees complain, he asks them to design their ideal job description in writing. He and they then discuss the job and look for ways to adjust the position to fulfill more of their needs and desires.
2. Planning reduces inefficient use of resources. Supervisors should spend a good deal of their time planning. A well-planned environment improves morale.
3. Getting employees involved starts by having employees help design their work environment. When presented with problems needing solutions, King challenges his people always to have an answer when giving a problem.
4. Keeping a fresh approach is essential to continue enjoying your job. If you can't find a valid reason to keep doing something, try stopping it. If that creates a problem, reconsider the situation. King likes to give people enough information; he also reminds them of the organization's priorities to make their own decisions.

Leadership is the key to a productive and motivated workgroup. To identify the elements of maintenance leadership, we discussed the issue with over 100 maintenance managers, maintenance supervisors, maintenance planners, plant engineers, building managers, and production managers throughout the United States and Canada. The organizations ranged from the largest industrial firms, federal and local governments, to small industrial and building management firms.

Supervision in a Union Environment

The manager must have leadership in a union shop. The same leadership skills are essential when there is a union involved. In addition to the skills already mentioned, other skills are particularly necessary for conflict, or when listening to a grievance.

But no matter if there is a union or not, specific skills are called for as a simple matter of courtesy:

1. Don't rush the discussion.
2. Active listening is essential. Try to suspend your judgments or experience, and listen to what the problem is like from the other person's point of view. You do not have to agree; just listen.
3. Set a time and place for the discussion. Show up on time.
4. Listen to the entire story before the discussion.
5. Follow through on the steps outlined in the bargaining agreement. If you've agreed to any action, follow through on it.
6. Realize that this is important to the person who brought it up, and treat it that way.

ACTIVE SUPERVISION VERSUS PASSIVE SUPERVISION

The new tendency is to fill up your supervisor's day with paperwork. Or if I ask the question: "How much time does your supervisor spend on the shop floor?" what would you answer?

I think too few people have put a value on the supervisor supervising! An active supervisor contributes in many areas. We call this active supervision. The question is, what is active supervision?

Active supervision is where the supervisor spends substantial time on the shop floor, helping workers solve problems. As strange as it might sound, on the psychological level, the supervisor might have to be both mother (nurturing and supportive) and father (strict and tough) to members of the crew.

Active supervision has several dimensions.

- *Ongoing performance monitoring*: The supervisor knows how long each job should take and checks it periodically throughout the day. A four-hour job issued in the morning should be done by the lunch break. When the jobs fall behind, the experienced supervisor thinks about the best intervention. In some cases, it might be logistical help, tool help, information about how to proceed, in some cases, a kick in the worker's pants! In other cases, the supervisor will hang back if wrestling with the job is vital for training.
- *Friend*: Everyone needs a friend. We like to think that the supervisor has our back. Sometimes the best thing a supervisor can do is to just listen like a friend. It can make all the difference.
- *Coach*: You are not on the field you are on the sidelines. Your job is to get the crew to win the game without picking up a tool.
- *Paperwork compliance officer*: The work order drives the accuracy of all analyses. If the work order is complete and accurate, then decision-making and root cause analysis is dramatically more straightforward. The supervisor is always auditing paperwork and returning it when it is deficient. He or she should always look at work orders on the floor and ensure entries are being made contemporaneously (at the same time) as the activity.

- *PM anal compulsive*: Do the PM as it is written. A related issue is PM compliance. If a worker does not have the task list in-hand when they are doing the PM, how do we know the PM was done as designed? The supervisor ensures the task list is carried with while on PMs.
- *Teacher, Mentor*: The supervisor should either be continuously training or directing the training of members of the crew. Everyone has areas that they are better and areas that they are worse. The easiest team to schedule is one where everyone can do everything. The effective supervisor should be reviewing the schedule every day and look for training opportunities. These can be formal training sessions or letting the trainee "help" an experienced hand.
- *Prison guard*: People break the rules. Some people text while working or do another dangerous activity. The supervisor needs to know what the rules are and enforce them. Sometimes a hard kick in the posterior is what is required.
- *Priest, Therapist*: People have problems. People bring their issues to work. The supervisor's job is to be sure the worker stays focused on his or her duties and that their problems are in the back seat during the workday. Sometimes it takes a therapist or priest to get people in the working frame of mind.
- *Quality control officer*: The supervisor is responsible for the overall quality of all work performed in his or her shop. Where there are issues with the quality of work, the supervisor determines the cause of the problem. The supervisor works with the worker to solve the quality problem. If the problem is with the company or system (such as adverse conditions, lack of tools, or parts), he/she should attack them too. If the worker has a problem with external issues, the supervisor should mentor them or find them some support in the organization.
- *Safety officer*: The supervisor should intervene any time an employee or visitor performs an unsafe act or is in the shop without the proper personal protective equipment. The supervisor is the champion for safety and makes sure the shop is safe.
- *Tidiness cheerleader (5S)*: The shop must be kept clean for safety, efficiency, and morale reasons. All clean-up for specific jobs should be part of and charged to the individual job. The supervisor should arrange for periodic clean projects to keep the whole area and the yard tidy.

2 Maintenance Mentoring and Role Models

There are many ways to develop as a person and as a supervisor. More often than not, the development happens simultaneously. Two good ideas are to look for positive role models and to seek out mentors. It is not a coincidence that there is an opportunity for development to become a role model and become a mentor.

DEMOGRAPHICS

We have a small window that is closing fast. The first baby boomers are 75, and the last ones will retire in only a few years. The last boomer was born in 1964 and will reach full retirement age (67) in 2021. In five years, almost the whole lot will be retired.

Our successors are taking over or are in training to take over. We want them to be successful (they are raising our grandkids, after all). It is our job, our responsibility, even our duty to accelerate the development of leadership. To the best of our ability, they will be ready.

The use of maintenance planning, PM, scheduling is slowly going away. It has already been going away for a decade or so, and it has become less visible. This is caused by retirements and the trends toward cost-cutting and over-optimization.

This upcoming generation is smart and tech-savvy. They are better prepared for automation and deep levels of computer control than we are. It is not primarily knowledge that we need to transmit, although there is some knowledge. It is not the skills we want to impart, although there are certain skills we wish to transfer. We need to transmit something of our attitude and something of our judgment and something of our pride in good products and on robust machines.

MENTORING

There are already classes in knowledge, skills, processes, and procedures. There is very little available about attitudes, judgment, and pride. One antidote is personal. Mentoring is a special relationship between a senior person (the mentor) dedicated to the business (and emotional) development of the person being mentored.

The story of Mentor comes from Homer's Odyssey. It is the story of Odysseus' (Ulysses) 10-year trip home after the fall of Troy. Telemachus was Odysseus' son. Mentor was Telemachus' teacher and advisor. The word mentor evolved to mean trusted advisor, friend, teacher, and wise person.

According to Emory University, Learning and Organizational Development: Mentoring is a fundamental form of human development where one person invests time, energy, and personal know-how in assisting the growth and ability of another person.

Mentors give advice. They might advise patience when that is needed or action when that is required.

There has been a lot of talk about the loss of knowledge and skills with these older folks gone, but I think this is less important than another loss because knowledge and skills are more easily replaced. This other loss is much less visible and potentially much more expensive.

What does it take to gain mastery of a craft? Sure, there is knowledge, experience, and there are skills, but mastery requires something else. All these three competencies (knowledge, skills, and some secret sauce mixed with experience) can be gained by yourself. But it is much quicker and surer if there is a master on or near the team. This person is a mentor and what they provide is intangible.

Sure, they pass on some skills and a spot of knowledge, but "the craft master's attitude" is the secret sauce. This includes where to (mentally) stand while solving a problem. It also consists of a certain kind of attention. We are now kind of self-absorbed. A master mechanic must be acutely focused on what the machine is saying and through a special kind of attention be able to commune with the machine.

A mentor models this behaver by not jumping to conclusions, by really paying attention to the sounds, smells, and sights of the machine. They also encourage the protégée that mistakes help them learn, that they should be serious but lighten up. A good mentor might be quiet and can act as a committed listener.

If you are on the youthful side of the divide, then a mentor can be useful for your career and accelerate your assentation to mastery. If you are on the far side of the gap, then mentoring can add a tremendous amount of satisfaction to your job and role.

The master offers lessons learned with the intent of helping the protégée learn how to listen, see, and think. Who is going to do this now?

I want to point out a book I just read that deals with some of this called Shop Class as Soulcraft: An Inquiry into the Value of Work by Matthew B. Crawford.

Some of the "things" the mentor might work on concern teaching the person to "see" and properly evaluate reality. They might provide context for what is going on at the company. There is also a big picture that might help the mentee understand a current crisis more clearly. One crucial role is to help the mentee be prepared and help them avoid being blindsided by events or people. Finally, the mentor can be a mirror so that the mentee can look and see themselves in the future.

The second big thing provided by the mentor is to listen deeply and powerfully to the mentee. Listening carefully and listening without filters or concerns (sometimes called active or innocent listening). The goal is to "get" the mentee's world view. "Getting" someone in this way will cause the person to get stronger and grow. This kind of listening is like putting fertilizer on a garden—it makes it grow.

The third "thing" is to show confidence in the person. Sometimes the only thing the person needs is for someone to believe in them. In the movie "Joy," the only person who believed in her (Joy) was her grandmother. In the face of universal negativity in her family, her grandmother's confidence was enough to carry her through.

There are some practical things the mentor can provide, including introductions to people who might be useful, resources for projects, help create and manage opportunities. A well-mentored person has an increased comfort level in a job or life. This comes from the enhanced relationship with reality. They tend to be more effective than they once were. One area that the increased effectiveness comes from is knowing what is essential and what is less essential. Finally, they have more fun at work, for the mentor, a feeling of satisfaction from contributing to someone's life.

SUPERVISOR IS A ROLE MODEL

> A role model is "someone others look to as a good example. A *role model* is worthy of imitation—like your beloved teacher or a well-behaved celebrity."
>
> (www.vocabulary.com › dictionary › role model)

Just as a model is a representation of something (like a model car or airplane), a role model represents someone who inspires others to imitate his or her good behavior. If someone misbehaves, you could say they're a negative or harmful role model—the kind of person who shouldn't be imitated.

There are roles in society. We learn how to act in and to people in these roles early on in our lives. Where do we learn to respect a judge, listen carefully to a doctor, or take our problems to a cleric? Some roles in themselves seem to create the condition of being a role model.

For almost everyone, their parents, older relatives, and older siblings were role models. The TV was a strong influence for role models in the last generation as social media is now. As you grow up, there are still role models. You also use military officers, teachers, and of course, people on the job as role models.

Supervisors would be role models even if they didn't sign up to be role models. People look to you, in some cases, also when you are younger than them, as a role model. This is a problem and creates some opportunities. The main problem is that your team members might be watching you more closely than you are comfortable with.

The opportunity is also that they are watching you closely. So, your actions model what you want from them.

Two basic types of role models

The positive role model: We think of these when we are discussing role models. They are the people who not only inspire you; they also guide you, motivate you to reach your full potential.

Negative role models: This is someone who you swore that you would never be like them. Negative role models are people whose footsteps you never want to follow.

For many people, the negative role models might have had the most effect on their supervisory style today. Their "badness" in whatever form caused you to make a solemn pledge to not be like them.

One last thought on your role models. If they are still in your life or you can even reach out to them, thank them. They were just trying to get through the day, and it might just make their day to hear what you learned from them.

3 Maintenance Supervisor Personality Profiles

Your people and your fellow supervisors come in many different flavors (personality types). Competent supervisors appreciate the differences and use the individual's traits to support the team and the mission.

The other aspect of personality types is that some people will rub you the wrong way. Their very existence is irritating. This simple assessment can show you that the problem is not personal to you but rather a function of their personality. Remember, sometimes the way they act it is not personal at all.

> The purpose of the Myers-Briggs Type Indicator® (MBTI®) personality inventory is to make the theory of psychological types described by C.G. Jung understandable and useful in people's lives. The essence of the theory is that much seemingly random variation in the behavior is quite orderly and consistent, being due to basic differences in the ways individuals prefer to use their perception and judgment.
>
> (Meyers-Briggs Foundation)

The original MBTI was developed by a mother–daughter team (Isabel Briggs Myers and her mother, Katharine Briggs) to make the insights of type theory accessible to individuals and groups.

The author researched maintenance supervisor personality styles by collecting and analyzing the results of about 500 surveys over two years. We used a shortened test by R. Craig Hogan and David W. Champagne.

Some of the background about the traits is from *The 1980 Handbook for Group Facilitators*, edited by J. Williams Pfeiffer and John E. Jones (San Diego, CA, University Associates, Inc., 1980). The author's introduction to the test was at a course at Boston University taught by Bill Ronco. Many of the explanations are from Bill Ronco.

To take the test: mbtionline.com
For more information: https://www.myersbriggs.org

According to the Meyers-Briggs Type Indicator, there are four dimensions of personality. Each dimension has two poles or traits that define it. Each of the eight characteristics has a letter designation. The letters stand for:

Dimension: I introverted–E extroverted

Favorite world: Do you prefer to focus on the outer world or your inner world?
Dimension: N intuitive–S sensor

Information: Do you prefer to focus on the basic information you take in, or do you prefer to interpret and add meaning?

Dimension: T thinker–F feeler

Decisions: When making decisions, do you prefer to first look at logic and consistency or first look at the people and particular circumstances?

Dimension: P perceiver–J judger

Structure: In dealing with the outside world, do you prefer to get things decided, or do you prefer to stay open to new information and options?

We wanted to see what types of people were attracted to maintenance. We asked over 500 maintenance supervisors and managers to take this test. The first percentage was its appearance in the general population, and the second percentage was the occurrence in the maintenance population.

General guidelines for the eight traits: The comment before each dimension describes the broad category; the percentage after each type is the percent in the general population, and the second percentage is the prevalence in the maintenance population.

I-E How Do You Choose to Relate to the World

I Introverted: 25% general population, 51% maintenance population. People who are more introverted than extroverted tend to make decisions somewhat independent from the rewards and constraints of the situation, culture, people, or things around them. They are quiet and diligent at working alone and socially reserved. They may dislike being interrupted while working. They also tend to forget names and faces.

Possible strengths	Possible weaknesses
Independent	misunderstands the external
Works alone	avoids others
Is diligent	is secretive
Works with ideas	is misunderstood by others
Is careful of generalizations	needs a quiet environment to work
Is careful before acting	dislikes being interrupted

E Extroverted: 75% general population, 49% maintenance population. Extroverted people are attuned to the culture, people, and things around them and tend to make decisions congruent with demands and expectations. The extrovert is outgoing, socially free, interested in variety, and in working with people. The extrovert may become impatient with long, slow tasks and does not mind being interrupted by people.

Possible strengths	Possible weaknesses
Understands the external	has less independence
Interacts with others	does not work without people
Is open	needs change and variety
Acts, does	is impulsive
Is well understood	is impatient with routine

N-S How Do You See the World

N Intuitive: 25% general population, 43% maintenance population. The intuitive person prefers possibilities, theories, the big picture, invention, the new and becomes bored with nitty-gritty details, the concrete and actual, and facts unrelated to concepts. The intuitive person thinks and discusses with spontaneous leaps of intuition that may leave out or neglect details. Problem solving comes easily for this type of person, although there is a tendency to make errors of fact.

Possible strengths	Possible weaknesses
Sees possibilities	is inattentive to detail, precision
Sees gestalts (holistic view)	is inattentive to the actual and practical
Imagines, intuits	is impatient with the tedious
Works out new ideas	leaves things out in leaps of logic
Works with the complicated	loses sight of the here and now
Solves novel problems	jumps to conclusions

S Sensor: 75% general population, 57% maintenance population. The sensing type of person prefers the concrete, real, factual, structured, tangible, here-and-now, and becomes impatient with theory, the abstract, and mistrusts intuition. The sensing person thinks in careful detail, remembering real facts, making few errors of fact, but possibly missing a conception of the overall.

Possible strengths	Possible weaknesses
Attends to detail	does not see possibilities
Is practical	loses the overall in the detail
Has memory for fact	mistrusts intuition
Works with tedious detail	does not work with the new
Is patient	is frustrated with the complicated
Is careful, systematic	prefers not to imagine future

F-T Which Type of Decision-Making Is More Comfortable

F Feeler: 50% general population, 47% maintenance population. The feeler makes judgments about life, people, occurrences, and things based on empathy, warmth, and personal values. Consequently, feelers are more interested in people and feelings than in impersonal logic, analysis, and things, and in conciliation and harmony more than in being on top or achieving personal goals. The feeler gets along well with people.

Possible strengths	Possible weaknesses
Considers other's feelings	is not guided by logic, not objective
Understands needs, values, feelings	is uncritical, overly accepting
Is interested in conciliation	is less organized
Persuades arouses	bases justice on feelings

T Thinker: 50% general population, 52% maintenance population. The thinker makes judgments about life, people, occurrences, and things based on logic, analysis, and evidence, avoiding the irrationality of making decisions based on feelings and values. As a result, the thinker is more interested in reasoning, analysis, and valid conclusions than in empathy, values, and personal warmth. The thinker may step on others' feelings and needs without realizing it, neglecting to take into consideration the values of others.

Possible strengths	Possible weaknesses
Is logical, analytical	does not notice peoples' feelings
Is objective	misunderstanding other's values
Is organized	is uninterested in conciliation
Has critical ability	does not show feelings
Is just	shows less mercy
Stands firm	is uninterested in persuading

P-J How Do You Handle Time

P Perceiver: 40% general population, 22% maintenance population. The perceiver is a gatherer, always wanting to know more before deciding, holding off decisions and judgments. Therefore, the perceiver is open, flexible, adaptive, non-judgmental, able to see and appreciate all sides of an issue, always welcoming new perspectives and new information about issues. However, perceivers are also difficult to pin down. They may be indecisive and noncommittal, becoming involved in so many tasks that do not reach closure that they might become frustrated at times. Even when they finish tasks, perceivers will tend to look back at them and wonder whether they are satisfactory or could have been done another way. The perceiver wishes to roll with life rather than change it.

Possible strengths	Possible weaknesses
Compromises	is indecisive
Sees all sides of issues	does not plan
Is flexible, adaptable	has no order
Remains open for change	does not control circumstances
Decides based on all data	is easily distracted from tasks
Is not judgmental	does not finish projects

J Judger: 60% general population, 78% maintenance population. The judger is decisive, firm, and sure, setting goals and sticking to them. The judger wants to close the books, make decisions, and get on to the next project. When a project does not yet have closure, judges will leave it behind, go on to new tasks and not look back.

Possible strengths	Possible weaknesses
Decides	is unyielding, stubborn
Plans	is inflexible, unadaptable
Orders	decides with insufficient data
Controls	is judgmental
Makes quick decisions	task or plan controlled
Remains with a job	wishes not to interrupt work

KEYS TO THIS PERSONALITY INVENTORY

- People who have similar strengths seem to click together; they also arrive at collective decisions more quickly. They seem to operate on the same wavelength.
- The problem with people who have the same strengths in the above dimensions is that they may reinforce each other's blind spots. Their decisions may suffer because they have similar weaknesses.
- People with different strengths may not see eye to eye on many things. The more the group differs, the more likely misunderstandings and conflict will occur.
- The advantage of groups with different strengths is that decisions may be improved as a result of the differing points of view.
- People might be sensitive about criticism regarding their areas of weakness. People often avoid using their weaker sides, and conflict might occur when they are pushed into using those dimensions or when others point out deficiencies.
- A person's character cannot be changed to its opposite. However, each person can learn where their strengths lie and figure out strategies to make greater use of them; they can also learn where their weaknesses lie and work to strengthen them.
- People's values, beliefs, decisions, and actions will be profoundly influenced by all four of their stronger dimensions.
- Groups with a preponderance of members with similar strengths should seek out and listen to other types of people when making decisions.

SUPERVISOR PERSONALITY DESCRIPTIONS

Maintenance supervisor personality profiles based on the 450 tests. Description of the 16 supervisor personality combinations and their frequency in the maintenance population:

ISTJ 14.5%

ISTJ is a maintenance person that is well organized, serious, and quiet. Respect for and facility with facts. On the surface, they are calm in a crisis but might have a vivid reaction underneath. They take responsibility for mistakes. Will make up their mind about a situation and work slowly, methodically regardless of protests or distractions. They tend to stabilize projects and workgroups. They are at risk of thinking everyone is like them and overriding less forceful people. They might also minimize the value of imagination and intuition.

ESTJ 13.8%

Very organized and likes to set goals. Practical and realistic maintenance person and matter of fact. More interested in the here and now. A naturally good mechanic. Like to organize and run things using logical processes. Natural maintenance supervisor when they can consider others' feelings and values. May decide too quickly.

INFJ 10.2%

This supervisor will succeed through perseverance and originality. INFJ is an influential idea person driven by inspiration. Care should be taken to avoid being smothered by routine aspects of the job. They put his or her best efforts into work. Quietly forceful, conscientious, concerned about others. Respected for firm principles. Desire to do what is needed or wanted. Must temper inspirations with the development of their thought process.

ENTJ 10%

This supervisor is a hearty, frank, decisive, well-informed leader. Enjoy long-range planning and thinking ahead. Good at logic, intelligence, reasonableness, speaking. The main interest is in seeing possibilities beyond the present. Maybe overly optimistic about the situation. ENTJ needs someone around with common sense. May decide too quickly and may ignore others' values and feelings.

ENFJ 8.5%

The supervisor is friendly, accessible, and responsive to both praise and criticism. He or she is persevering, orderly, and conscientious. They are interested in possibilities and harmony. Concerned about what others think, want, and feel. Can present proposals or facilitate groups with ease and tact. Might tend to jump to conclusions. Have many rules, shoulds, and should not.

ESFJ 8.1%

ESFJ describes a warm-hearted, outgoing maintenance supervisor who is a born cooperator. Little interest in technical areas. Will create harmony and find value in other people's opinions. Practical, orderly, down to earth, works best in a warm, encouraging, supportive environment. The ESFJ has a lot of shoulds, rules, and assumptions about situations.

ISFJ 7.9%

Quiet, friendly, responsive, and dependable supervisor. Will work devotedly to meet an obligation. Will lend a stable influence on any team. The ISFJ cares about the people on their crews. Thorough, painstakingly accurate. Needs time to master technical areas. He or she is at the risk of becoming withdrawn. Also does not trust imagination and intuition

INTJ 4.2%

INTJ Maintenance supervisors have original minds and offer great ideas. Strong intuition. Power to organize and carry out jobs with or without help. Will drive others as actively as themselves. Also, they can be considered stubborn, skeptical, and independent. They are single-minded and should seek the input of others to balance them. They need to develop their thinking so that they can evaluate their inspirations.

ISFP 4%

A quiet, friendly, sensitive, modest, good type of maintenance supervisor. Might not show warmth until they know you well. Looks at the world through their own deeply held values. The ISFP doesn't like disagreements. Craft oriented. Loyal followers. Do not usually choose to lead. Likes a slow pace. Might not measure up to inner ideals and feel inadequate.

ESFP 3.6%

ESFP is a supervisor that is outgoing, easy-going, accepting, and friendly. Very flexible problem solver not bound by current rules and procedures. Actively curious about people, objects, food, or anything sensory. Knows what is going on through the grapevine and participates. Remembers facts better than theory. Learns by doing. Has good common sense. They might be too easy on discipline. Their love for a good time might put them at risk.

INTP 3.6%

These quiet maintenance supervisors are extremely logical, abstract, and interested in ideas. They have sharply defined interests. More interested in ideas than the practical application of the concepts in the world. Little need for social small-talk, get together, and anything else not in their area of interest. Might overlook other people's values and feelings.

INFP 3.4%

Enthusiastic maintenance supervisor, little concern for surroundings or trappings. They are driven by inner convictions that are usually difficult to express. Enjoys learning, ideas, languages, and independent projects. They can be very absorbed in the current activity. Sometimes excellent writers. Not usually talkative until they know you well. May feel they don't measure up to their internal standards.

ISTP 3.2%

Cool, quiet, and a reserved supervisor who analyzes everything with detachment. Prefers to organize ideas and facts rather than people or situations. They are interested in impersonal principles and why mechanical things work. Could have a high capacity to understand the facts of a situation. Engineering-oriented. Will not overexert themselves. May overlook other people's needs and values. May not follow through.

ENTP 1.7%

The ENTP is a supervisor who is quick, ingenious, energetic, and good at many things. Their thinking helps temper their intuition. Might neglect routine assignments. Resourceful at new, unique, novel problems. Stimulating company and might argue either side of an issue for fun. They must always feel challenged. Needs to learn to follow through. Without the development of their judgment, they will waste their energy on ill-chosen projects.

ENFP 1.5%

The ENFP is a warm, enthusiastic maintenance supervisor. Can-do attitude; sees possibility everywhere. Quick to help and with solutions. Experienced people handlers with great insight. Will improvise rather than plan in advance. They hate routine. They might leave projects after the core problems are solved. They become bored and leave projects incomplete.

ESTP 1.2%

No worry, no hurry, supervisor. They operate from concrete reality—what they can see, hear, touch, taste, or smell. They enjoy sensory pleasures. Not bound by current rules and procedures to find the solutions to problems. Like mechanical things and movement. Blunt and occasionally insensitive. Natural craftsperson when they can include people. Even after they are supervisors, they keep a mechanical hobby at home for the joy of it. Their love of a good time might be a risk.

The test is designed to measure a person's personality traits in Jungian terms (Carl Jung first described the qualities in 1921). Others have worked on it, including Myers (1962) and Hogan (1979).

If you are interested in detailed descriptions of the 16 types, write to B & D Book Company, 1400 W. 13th Sp. 128, Upland, CA. 91786 for excerpts from *Please Understand Me: Character and Temperament Types.*

4 Company Culture

"Culture always eats strategy for lunch!"

(Peter Drucker)

Whatever you are worried about, it is not you, it is your culture!

At a new job, if you look around, all you see is culture. The culture is in the tools, housekeeping, how people talk to each other, what people complain about, it is all culture.

Regardless of how good your reliability strategy is, it is your organizational culture that will determine its performance. Culture is built from within, and it is "cultivated" by leaders who aim at engaging employees in delivering reliability and asset performance of the organization.

WHAT IS YOUR CULTURE OF RELIABILITY?

If you take a great worker and put them into a nasty environment within maybe three months, they will look like the people around them (perhaps a bit better). If you take a lousy worker and put them in an excellent environment in three months, they will look like the people around them (maybe a bit worse). It is all about culture!

One company has a culture that is famous for their orientation toward safety.

For example, why do you suppose DuPont became so serious about safety?

- First business—making gunpowder and explosives
- Founder knew the risks and ruled:
  - No employee may enter a new or rebuilt mill until a member of top management has personally operated it.
  - Du Pont raised seven children on the site, and on at least one occasion, his wife was injured, and the house was severely damaged in an explosion.
  - Ultimately, several family members died of work-related causes.
  - First fatality in 1815: 9 deaths
  - 1818: 34 deaths

He created a culture of safety, no joke! It survived over 200 years.

ORGANIZATION MISSION, VISION, AND VALUES

All organizations have a purpose for being. Someone or some group thought it would be a good idea to create an organization for some specific reason. This statement could be from a single sentence to a long document. This is the reason for being of the organization. It encompasses the mission, vision, and values of the organization. It is the reason the organization exists.

Always operate in alignment with the organization's mission, vision, and values

This statement tells you, without a doubt, what to do in almost all circumstances. It provides the context or story of why you are doing what you are doing.

The organization's leadership and patterns drive culture. Culture and reliability are always a match.

Your culture is the sum of the customs, traditions, social institutions, and achievements of a time in history, country, company, or other social groups.

Some of the factors that contribute to your culture can be examined directly by thinking about some questions:

Leadership

- Does leadership do what they say and say what they do?
- Comments about the organizational commitment to integrity internally and externally
- How well is your company run from its openness to improvement?
- Is leadership focused on a short term or a balance of short and long term?
- What is the attitude toward a clean shop?
- Is safety a real priority?
- What is the political situation at home (top leaders)?

FIGURE 4.1 What does this cabinet say about culture? What about the sign?

Business

- What business are you in popularly, size, profitability, stability?
- What assets? Special hazards of equipment, process, product, or location?
- What failure modes in your use or geography or any particularities in terrain?
- How good are your engineering skills, tools, technology?
- Describe the qualities of the workers' skill levels, training, goodwill.

You see, breathe, smell, and hear the culture everywhere and in all activities.

FIGURE 4.2 What does this workbench say about the culture?

- What do you see about the culture of this workbench?
- From this panel?

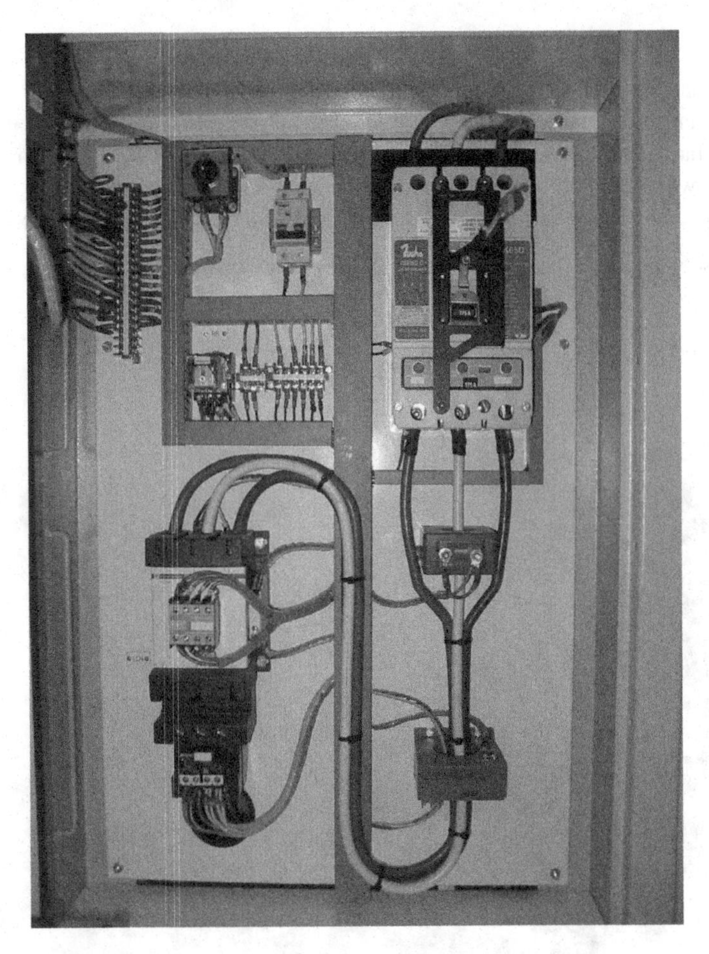

FIGURE 4.3 What does this panel say about the culture?

5 Getting Through to Other Humans

Your job is to get through to other people. Despite the noise, language barriers, prejudice, cultural differences, it is your job to communicate and to ensure the communication is received and understood. Lives may depend on it.

COMMUNICATIONS

Basic communication

- Communications support management objectives
- Excellent communication is a skill that can be learned
- You will feel better about yourself if you are a better communicator
- Life skill is also useful in other areas outside work

Communication requires:

1. Some discipline
2. Techniques
3. Awareness and attention
4. Reasons to communicate

Did you know that 70% of the message of communication comes across via non-verbal channels? Things like:

Voice tone	Voice volume	Body positioning
Facial expression	Animation	Air space
Eye contact	Fluency	Physical barriers
Hand gestures	Skin coloration	Breathing patterns

These channels carry your messages

Important Issue

Note that cross-cultural, cross-ethnic, cross-sexual, and cross-class communication differences can lead to misunderstandings.

When dealing with people from different groups than yours, be aware of assumptions, stereotypes, golden rule thinking, unintended meanings, different gestures, and different customs.

Travelers are told to learn about the culture they are visiting. Keep in mind that when a foreign-born person joins your team, some of these faux pas are still unconsciously present and effecting their views.

See Appendix for specific behaviors for international travelers.

6 Motivation

You want to be the kind of supervisor who motivates the workers to perform excellent maintenance work. The way you motivate them is vitally important. Once you understand the mechanics of motivation, you will be able to choose the way you motivate your team.

Motivating maintenance workers can be tough. Having a motivated workforce is not an accident. It is a combination of a deliberate approach with action. There is little mystery to it.

Ron Augustine, a supervisor for Parker Hannifin Corp. in Michigan, says, "For ten years my supervisor took credit for the things that we did that went right, but when things went wrong he was the first to point the finger." Would you feel motivated to do anything beyond the minimum, working for such a man?

How would you like to go to work every day like David Daugherty for 1615 L Associates in Washington, "It's an environment where everyone does their own thing, and no one tries to work together as a team? Employers don't seem to care about what happens. The building systems are neglected; they fall apart, and we rush to fix them." That sounds depressing.

At Eastalco Aluminum in Maryland, Roy Ellison supervises a crew that just went through an 81-day strike. The result is $0.92 per hour cut after no pay increases for the last six years. He reports, "The number one problem is morale and motivation."

We can all identify with these situations. Note that in the first case, the problem was with a person, in the second with a department and the third with the entire company. Ideally, motivation is the result of the three levels working well together in concert.

You can have a motivated workgroup despite a negative department or adverse company situation. It's just harder to make it happen and then maintain it.

THE PROBABLE STRUCTURE OF MOTIVATION

Much work has been done in the field of motivation for all types of workers. Any level of success in this area will ensure a researcher's limitless funding and opportunity. Right now, there is no overall theory of motivation supported by both scientific study and practical application.

One of the classic theories of motivation comes from psychologist Abraham Maslow. While his arguments are widely known, there is very little scientific support for them. Yet Maslow's theories, with some modification, closely correspond with what we observe in the stories that follow.

Maslow taught that there are five categories of needs motivating people. These categories are loosely organized into a hierarchy.

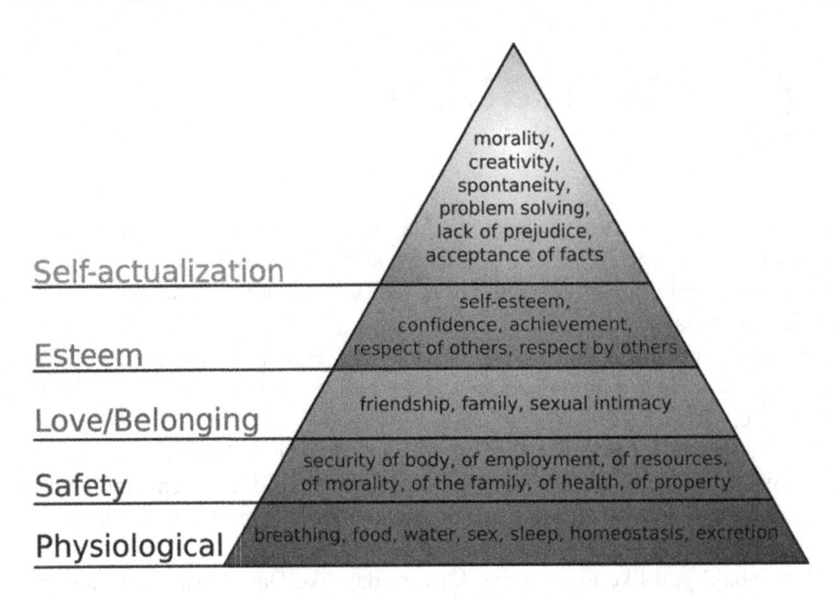

FIGURE 6.1 Maslow's hierarchy of needs.

When the lower-level needs are unfulfilled, they tend to assert themselves and become dominant. This is easiest to see with the lower-level needs: having friends is important but rapidly becomes less critical when you fear for your safety or can't breathe. Everyone has some level of satisfaction within the hierarchy. As you review the stories below, you can use Maslow's hierarchy to evaluate how each person or group is being motivated.

Let's examine some significant themes for successful motivation. There are two types of activities and attitudes described below. The first is a daily routine for conducting business; the second concerns one-time (or occasional) interventions. Cases come directly from currently active maintenance supervisors:

Type I: 8 Everyday Motivational Techniques

Recognition is by far the easiest way to increase the motivation level of most maintenance workers. Recognition comes in many forms. For example, Johnny Johnson of Abex Corp. in Virginia reports:

> We had a press that needed to be converted to run a new product in a short time. I acted as a cheerleader and got all three shifts working together. The job got completed in 1/3 the normal time. I wrote a letter of appreciation for each person for the personnel file. That motivated them.

In another case, recognition was extended to include the family. Gabriel Saavedra of Baxter Hyland in California once had this motivating experience:

> Upon completion of the job, the Vice President sent me a note of thanks with copies to my boss and my file. He also sent a basket of fruit with a note of explanation to my wife.

Why would anyone work hard if no one ever finds out about it? But many, possibly even most people are willing to go the extra mile if people recognize what they do. Acknowledge your people for what they do, and you earn their loyalty and their best efforts.

Treat them with respect: I think this goes without saying. Bernard Rulle, a carpentry shop foreman at George Air Force Base, has some rules for himself:

- I like to know my people personally. I know their hobbies.
- I listen to people with my mouth shut.
- I remember their birthdays and send cards.
- I praise them in front of their peers when they do outstanding work.
- I ask for donations when loved ones die for plane fares or expenses.
- I invite people without families to share Thanksgiving with me.
- I have a BBQ for the whole shop once a year.

Training and ongoing development: You have a choice. Train your people and keep them up to date, or let it go and eventually find a sloppy, uncaring, unmotivated workforce. Factories, equipment, and facilities are changing, and our bank of skills is not changing rapidly enough to keep up. We must train people in new technologies and, for maximum flexibility, cross-train in two or more crafts. At Chevron Canada in Burnby, British Columbia, Michael Edwards had a problem with the computer process units.

> The computer-controlled process units were maintained by instrument technicians and process engineers. The technicians cared for the hardware, and the engineers looked after the software. The systems did not work well because there was not good communication between the two groups. We trained the technicians to write and maintain the process software. Since they had complete responsibility, they worked harder, and now the system runs smoothly. The instrument technicians are confident and highly motivated.

Give them real power: When you give a crew the ability to do the job their way, you'd better prepare for change. A highly motivated team with the power to make improvements can be very effective, as Edward Wilanowski from CIL Inc in Alberta Canada, testifies:

> Reactor rebuilds took 36 hours. We needed to reduce this time to increase productivity. Management tried increasing the number of people, but that increased non-productive time while people waited for each other to complete jobs. The extra labor did decrease the turn-around time a small amount. Management then tried pre-planning, which also slightly decreased the turn-around. Then the power was given to the tradespeople. They changed some equipment, added some tooling, and now rebuild a reactor in 16 hours with the original crew. They are a highly motivated crew.

In another example back across the border, Paul Redding, a supervisor for Purdy Corp., had a problem. He took over the Maintenance and Housekeeping Department. The group had a terrible reputation. They were known to be lazy and unproductive.

Office personnel mainly complained about the janitors. Paul called a meeting which went like this:

> At first, there was silence. After a long while, someone broke the ice with complaints about the cleaning materials. The flood gates opened, and I found out about several problems with the supplies we were purchasing. After the meeting, I gave them the power to purchase their supplies. They were given a monthly budget and the ability to make their own decisions. Our purchasing department gave a mini-course in buying. The group is now highly motivated, and complaints have dropped dramatically.

Ownership is a key concept from the 80s and 90s. We assume that someone who owns something will treat it better than someone who is merely a cog in the wheel of production. The worker, as the owner has a say, from acquisition through maintenance until their retirement. At Kimberly-Clark, Patrick McDonough wanted to put this into action:

> We had to purchase a new, expensive piece of equipment for the mill. Normally I assigned an engineer to study the need and develop a specification for the new unit. In this case, we put together a team with people from engineering, production, and maintenance. They worked on a specification together. Unlike other purchases from an "aloof" engineer, this equipment was installed with minimum trouble and was accepted quickly by all parties.

Planning: Many people enjoy working in a planned environment. A prepared environment has fewer surprises, a more smooth workflow, and higher productivity overall. One of the most important things to plan is your annual shut-down. Keith Brown from Dow Chemical describes the improvement to morale made by planning and excellent communication of the plan:

> The first time we did a shut-down with pre-planning and scheduling was an eye-opener for the crews. They enjoyed knowing their next jobs and not getting in each other's way. The people liked the visible schedule board.

Use their ideas: Frequently, the best ideas come from the people closest to the action. Skilled maintenance workers have significant experience that, once unleashed, can solve major and minor problems. When maintenance supports production, meaningful new ideas will flow. Ron Howard works for the Rice Growers Association of California, and had a problem with the low output from a packer:

> The supervisor got everyone together. He stated the problem of low output and suggested a few of his ideas. He then backed off and let the maintenance people come up with their ideas. In the end, he turned the project over to them and helped the effort by purchasing the parts and materials. The result surprised even him; the packer was hitting production levels 30% higher than factory specs and double the old output.

Communication: When people understand each other, they are more likely to work together and enjoy higher morale. High morale translates into smoother functioning with open channels of communication. Jim Wakefield, a Weyerhaeuser, maintenance supervisor, had poor communications between shifts. He solved the problem in a straightforward way:

> We set-up an inter-shift meeting to discuss all of the issues that affected the workers. The first few meetings seemed worthless. As time went on, more and more things came out. Now, if someone is on the job when the shift ends, they will stay over a few minutes and bring the new person up to speed.

Type II: 6 Intervention Techniques

The boss is one of us: In selected settings, letting people see the reality of the boss as a person can be an excellent intervention. Joe Costa, a lead mechanic at Angus Biotech Inc. in California, felt motivated when one of the bosses took off his suit and got down in the ditch. The story went like this:

> It was a hot, sweaty, filthy job that summer. We were working 12-hour shifts, seven days a week, to complete the shut-down work. Upper management would walk by and criticize this or that but never got too close. The chief engineer (who was on temporary assignment, supervising maintenance until a manager could be hired) decided to jump in and work with us. He supervised from the inside, took his meals with us, and worked shoulder to shoulder with the men. Morale skyrocketed. Our whole view of top management shifted. He showed that he was a regular person and could keep up with us to boot.

Challenge and competition: People love a game. Challenge ("I bet you can't wire the new grinder by Friday") is a great way to complete projects. Competition between groups can be useful and enlivening for the workgroup. William Castro, the maintenance manager of Ibis Systems, set up the following challenge:

> I gave each person an area of responsibility. We developed a Trouble/Failure report that indicates the amount and reason for downtime. Each area competed for the lowest downtime statistics. People started to take an active interest in the areas after the competition started.

Sometimes the challenge is against bringing in an outside contractor. By bringing in outside expertise, the maintenance workforce is told, in so many words, they are not trusted for this critical job. Charles Jones, a first-class maintenance mechanic for E-Systems Melpar Division in Virginia, relates an unintentional challenge that was an effective motivator:

> Early last summer, our 450-ton chiller went down because of the motor. The manager panicked and immediately wanted to call a contractor. I convinced them that we had the skills to do a good job in less time at substantial savings. I told our people that management didn't think we could do the job. We rebuilt the motor and had the chiller back online within one day at significant savings.

Drama: Sometimes, when there is a problem, the supervisor needs to "hit everyone over the head with a 2 × 4," through the judicious use of drama. Continental Mills in Seattle had such a problem with wasted product. The plant engineer, Dave Sloan, related this story:

> The president called a plant meeting and explained that we were losing $45,000 worth of product each month. He opened a case and proceeded to dump 45,000 one-dollar bills into a garbage can while saying, "This is what we are doing with our product." The impact was terrific. Our waste dropped 50% the next month and has continued to improve since then.

Reassignment: Even a single abrasive personality can significantly undermine motivation within a workgroup. Frequently reassignment, rotation, or transfer of the disgruntled parties will restore the workgroup's motivation level. Ron Vanderpool worked with a shop steward whose nit-picking attitude affected the crew. The organization was also suffering from quality problems. Ron suggested to management:

> Appoint an hourly employee to the quality program at the finish end of the process" (where the steward was senior). He then sold the program to the steward, who appreciated this type of input. Quality went up sharply while the steward got to nitpick in a way that helped the organization.

Survival is a powerful motivator. When you face a survival situation, people will work very hard to solve the problems. When faced with certain lay-off, C. Bostic of Hercules, Inc., went to management to save his people's jobs:

> We formed a 7-person central day crew. This crew would do construction jobs presently being done by contractors. Management bought the idea and started the project the very next week. This was a win–win situation. The supervisor solved the lay-off problem, and management looked good to the workers.

Variety: People get bored. Sometimes low motivation levels are the result of work that underutilizes people's capabilities. Marvin Barry from Ocean Construction Supplies in Maple Ridge Canada remembers when:

> Our mixer men were becoming tired, missing lubrication points, and missing inspections. To create more interest, we rotated the people through several parts of the plant and some nearby plants. The rotations seemed to spark enough interest so that the work was done well, and the people seemed happier.

7 Goals Are a Motivational Technique

Goal setting and goal management are dandy motivators because they appeal to people to people on many levels, and they can serve to get jobs done in maintenance. Consider using goals for PM projects, Defect Elimination projects, training, and behavior modification.

One way to motivate people is to have them set a goal. Goals appeal to the esteem and self-actualization levels of Maslow's hierarchy. People will exert themselves heroically (sometimes) to achieve a goal of a game or an activity that they have agreed to carry out. Whether that goal is a personal best, world record, or just finishing a paper, it doesn't matter; people will push themselves to get over the finish line they set.

Goals are the desired result; a person envisions plans and commits to achieve a personal or organizational desired endpoint. This is a fancy way of saying people like to have a target, and they will work harder to make it. Some of the great things about goals are that they call people to action. The default mode for many people is inaction. Goals change that and seem to add energy into the equation.

Goals and games are closely related. Games are fun. People compete to achieve a goal (sometimes literally). Goals make ordinary actions into a game. Lean maintenance projects can be tedious, but a lean maintenance game might motivate people to try harder (play harder).

If you think about it, sometimes people forget what is essential. If you have designed the crucial parts of your job into a series of goals, then the goals provide structure for "What should I be doing as a Leader?"

One definition of power is the ability to get things done. Guess what? Goals ramp up your power in whatever domain you are interested in. Think about the target as a magnet pulling you toward the outcome.

KEEPING GOALS ALIVE

Goals only have power when you remember you are in a game to achieve them. Just think about the power of your new year's resolution after a few weeks (or even a few days)? Usually, people forget they resolved to reach a weight or to call home every week. Dead and forgotten goals have no power. They may drain energy because they remind you of your powerlessness. So, forgotten goals = No power

Kinds of Goals

- Smart goals
- Stretch goals
- Impossible goals

The three types of goals have different functions. Smart goals are best to enhance personal productivity for the short term. Almost all literature about goal setting concerns these SMART type goals. Stretch goals are best used for the longer term. The person who undertakes stretch goals knows that at their current level of development, they are not up to the new higher purpose. They will have to stretch to get there. Impossible goals are guides for how one lives in the world. You will never achieve the goal, but your life working to get there will be meaningful to you and satisfying to live.

SMART Goals

SMART goals are great to accomplish some concrete outcome such as finishing a defect elimination project by March 1 or check ten parts for quantity on hand and see if the stock level is too high by the next meeting.

To make the goals even more powerful, making the goals into a game is motivating for many people. Smart goals (and games) have common characteristics

- Specific
- Measurable
- Attainable
- Realistic
- Time-bound

✓	SMART Goal Guide
	Specific: *What exactly will you accomplish? If your goal is extensive or comprehensive, break it down into smaller goals.*
	Description:
	Measurable: *How will you know when you've reached your goal or if you are making progress? Can you quantify the outcome?*
	Description:
	Attainable: *Is attaining the goal realistic with effort and commitment? Do you have the resources to achieve this goal? If not, how will you get them? Is making the goal dependent on anyone else? Is it possible to reframe it,so it only depends on you or your team? What factors may prevent you from accomplishing the goal?*
	Description:
	Relevant: *Why is achieving the goal important to you? Home in on why it matters. What effect will attain the goal have on your life or others?*
	Description:
	Time-bound: You can use any period, *but it is best to keep it under three months. If it's a large or complex goal, you may need to break it into smaller and more manageable increments with deadlines.*
	Description:

FIGURE 7.1 SMART goals guide.

Format for Smart Goals

If you are working with SMART goals, one way to effectively sabotage the process is to choose goals that sound great but are too global (we are going to be better maintenance people). Reasonable goals are specific goals (we will reduce breakdowns to less than two per year.)

Stretch Goals

The goal of this type might be to go to Graduate School or become a manager. When you talk about that goal, you might not be ready to undertake it. This because you don't know what or how to do the steps. You will have to stretch yourself and go places you've never gone before.

Stretch goals take belief in ones' self and usually a support network around them. A young band might see themselves at the Grammys. An entrepreneur might see the IPO (stock offering to the public).

- Go beyond what you know you can do or have done
- May take some thinking about how to do the goals
- May get you out of comfort zone
- Not necessarily easily attainable
- Causes the person to be someone different
- Involves growth and development
- Even failure can mean progress

Impossible Goals

The feeling of wellbeing might be related to dedication to a cause or idea bigger than yourself. During an interview about how it feels to play at 90 years old, Pablo Casals (who plays the cello) reports that he forgets his pain and age and exists in service to the music. When Mother Teresa was alive, she did not sweat the small stuff since her life is in service to a greater thing then herself.

Impossible goals create a way of life. People depend on you. When your life is about big goals, you don't sweat the small stuff, and it is all small stuff.

KEEP GOALS POWERFUL

Don't forget them. Keep your goals in existence! Stay in action toward the goal. Figure out what will be sufficient to get you off your duff and back into action.

- Display on wall
- Screensaver or banner
- Reminders in Outlook
- Teams
- Minutes and agendas

The trick is to keep them in existence (forgetting a goal takes it out of existence). Some of the ways you can keep them present are by putting up displays (posters), screensavers, banners that crawl across the screen, having pre-scheduled reminders in Outlook, or having teams.

There is a whole bunch of ways to keep a project, resolution, or idea in existence.

Pitfall: Noise!

Any structure you build to maintain your goals must break through the noise of your life, environment, and the on-going commentary in your head. This distraction can be as simple as the noise on the shop floor (so you can't hear the alarm you set on your phone) to complicated like your preoccupation with what is going on at home.

Noise is pervasive. Externally our society is noisy (just visit a restaurant, bar or almost any public venue). Less obvious is the inner dialog that people listen to. It consists of your opinions, self-criticism, gossip, thoughts, dialogs, rehearsals, dreams, fears, wants, and ideas.

People can be looking straight at you, and most of their listening is focused internally. Breaking through this noise is tough, and even when you breakthrough, you run the risk of getting blocked by a new round of internal noise. One element in your favor is that the person asked for the structure to help them with their noise.

Structures Breakthrough External and Internal Noise

These structures have some attributes:

- **Structures remind participants to do what they promised or were assigned by when they promised**
- Structures call you to be in action
- Structures live outside of your mind or memory (like the items above to ensure you don't forget)
- Are real (as opposed to a thought), physical (like a calendar) or virtual (Outlook)
- Designed to work within the boundaries of your life and support you in the fulfillment of your commitments
- Structures breakthrough the noise
- Structures can are either adequate or inadequate or as robust or weak

Examples of structures

Tricks to not forget or to keep things you speak in existence. The not-so-secret trick to success: You win if you stay in action—no matter what!

- Create reminders outside your memory
- Add people and tell people what you are doing
- Calendars (scheduled events)
- Add scheduled meetings
- Create mnemonic devices to remind you
- Add any media such as notes, signs, displays of progress, posters, forums, pictures, songs, videos that remind you
- Outlook reminders
- Bulletin boards
- One (sometimes overused) tool is to post the goals
- Postcards you mail to yourselves
- Even post-it notes are structures
- Add events such as rollouts, celebrations
- Add official recognition for progress

- Make the people teach others
- The goal is a conversation designed to keep you in action toward something.

Using Teams

Teams can be a structure to reinforce the existence of goals. The process of reporting progress to the team and discuss the goals' development is a powerful reminder. This tip works best when the goal meetings are regularly scheduled.

The specific elements of a weekly meeting that keeps goals in existence are the Minutes and the Agendas. The sessions can be used to get resources, ideas, and other help.

FAILING FORWARD: WHY GOALS WORK EVEN IF YOU FAIL

Of course, we want to achieve our goals, and we want our people to achieve their goals. The great thing about operating with goals is that they will pull us further even if we don't meet them. So, failure in a project will still leave you further along your path than not having the plan in the first place.

Last Word

The most successful people in the world use structures to remind them of what they promised and what was requested by subordinates, partners, and bosses.

Think about presidents, religious leaders, military generals, and others at the summit of their career. The thing they have in common is a series of structures to call them to succeed.

Lots of things to do with your time. Most of it will not impact what you are committed to, but must be done anyway.

It is essential to identify those activities that will "move the needle" closer to your goals and commitments.

8 Coping with Difficult People

As a supervisor, you are guaranteed to have to interact with difficult people. Many people think that there is a concentration of difficult people in or near maintenance. Building muscle in dealing with difficult people is a useful life skill that can work with friends and even family.

What are difficult people/situations?
We are going to deal with a subset of difficult people and situations. We are not going to discuss the difficulty of earthquakes, wars, pandemics, riots, explosions, car/airplane crashes, plant closings, shootings, and other outside disasters. We are going to deal with the effects of unrealistic production goals or hours, old machinery that always breaks down, employees who seem to cause problems, users who seem not to listen to reason.

How to detect the presence of a difficult person/situation?
There is one sure-fire method to determine if you face a difficult person/situation:

> Listen to your body. Your body will tell you through several, mostly unpleasant sensations. These sensations are the result of the activation of the primitive parts of your brain.

Millions of years ago, this primitive part of the brain prepared the pre-human to fight or for flight. Sensations include heart pounding, dry mouth, sweating, faintness, vision changes, clenched fists, changes in breathing, and a strong desire to act. This autonomic response is the key that a difficult situation/person has activated you.

Please note there is nothing wrong with this response. If your life were in danger, the reactions might save your life. The problem is that the average encounter with a problematic person is usually not life-threatening (even though it might feel that way).

Understanding your response to difficult people
There are things, situations, and personal attributes that irritate you. You will be irritated whenever they occur.

Exercise to determine some of the things that push your buttons

1. Try to identify some things that push your buttons (something that you have difficulty with).
2. Circle a few that influence you.
3. As you circle each one, remember the last situation of that type that irritated you. Try to visualize the person's face, hear their tone of voice, etc.

Examples:

Last-minute schedule changes
Hazy, unclear orders
Sarcastic people
Users/tenants that put you down
Being told you're stupid
Fidgety people
People that go behind your back
Wimpy bosses
Whiners
Foul-mouthed people
People who won't shut-up
People with excuses
People being absent
Suppliers that lie to you
Moody people
Angry bosses
Bullies
Companies with low moral standards
Employees that act like children on the job
Something that should work and doesn't
Emergencies that are not emergencies
Being called out of bed because a plug is pulled out
People who breathe down your neck when you're working
Being told, "We've already tried that, it will never work"

The difficulty of difficult people or situations exists within us
This idea may shock you, so be sure you are seated. The rule here is that the person suffering the distress is the one with the problem. As we have found, other people might not think that your situation is difficult at all. The issue is not how we are going to change the person/situation annoying us, but how we are going to cope. Changing the actual difficult person or condition would be great, but (sorry to inform you!) it doesn't happen frequently.

FIVE TECHNIQUES TO COPING WITH DIFFICULT PEOPLE/ SITUATIONS

The five techniques can be done sequentially or you can jump into the one method that is needed at the moment. The general rules of successful coping are the same as the rules for successful living. These skills make you more effective in dealing with all types of difficulties. Note that these are general rules, and many situations require a different or immediate response.

All methods of coping take practice. You won't necessarily be good at it right off. The key is to practice the new behavior.

The rule of repetition: Psychologists have determined that when you have a new behavior, it will feel fake until new pathways grow in your brain. How long it will take depends on many factors. One article said that it took 40 repetitions of new behavior for an adult (kids only need 20 repetitions) to establish new effective pathways.

1. Technique 1. Make Space for Yourself

The first rule of coping is to make space for yourself. Since the reactive part of your brain is activated, your ability to think is diminished. But if you withdraw from contact with the triggering situation, that will allow you to settle down. It will also enable the "difficult person" to settle down.

A. Try leaving. If you can't physically move, then
B. Try breathing slowly and deeply (only 5–10 connected breaths may be needed).
C. The stress reduction books say to spend 10 or 20 seconds visualizing yourself at your favorite vacation spot or doing your favorite activity.
D. Tell the person you are too angry to deal with them productively and make an appointment for later.

2. Put Yourself on the Map (PYOM)

Cultivate an expression of serious interest, if necessary, practice in a mirror.

Many times, the first item of business is to Put Yourself On the Map (PYOM). PYOM is a psychological stand that you take for yourself. PYOM is standing up for yourself as a person with rights. PTOM is not yelling louder than everyone else but asserting yourself.

There is an aspect of PYOM that is important and very difficult to understand. Everyone's personality is divided into parts. Different parts may be in control at different times without our realizing it. An example is when a challenging situation or person activates us. We shift from our everyday working personality to our more primitive activated behavior (which takes only a split second to make that change). At the moment of activation, we are prepared to fight or run. There are many gradations of this response.

PYOM activities start an internal dialogue between your everyday (higher) self to your activated (more primitive) self. The communication says something like, "Thank you for alerting me that there is a difficult situation at hand. I will protect us from harm. I will assert our rights. You can rest now, and I will protect us." This communication allows your activation level to drop and for your everyday self to be able to listen.

In some circumstances, PYOM activities seem courageous. Putting yourself on the map alerts the others involved that you are a reasonable, worthwhile person, and they cannot trample over you. Putting yourself on the map ensures that you are an equal.

PYOM statements include:

- What you saw or heard, with nothing added, and no judgment. ("I saw you go into the supply cabinet," not "You're always going into the supply cabinet, you liar.") Stick to observable (not inferential) and objective (not judgmental) material. Use "I" statements. Be specific, brief, and make sure it's directed to the right person.

- How do you feel? Describe your current feelings. No one, but you are responsible for your feelings. Don't add "I feel …. because you …." Stop with your expression of your feelings. Include the degree of feeling (I am mildly ___; this is the angriest, etc.).
- Describe the tangible effects without embellishment, for example, after production pulled your lube person to move material, say something like, "The result will be increased breakdowns because we don't have the manpower for the lube schedule anymore."
- Sometimes describing your understanding in a non-threatening way is enough to give the other person the space to calm down. Once both of you calm down, a resolution may present itself.
- Shift into listening by becoming silent. Begin active listening.

3. Listen to Your Truth

What about this interchange is about you? Think about how you are and how you could have contributed to the situation. Tell yourself the truth. If you are responsible for or helped contribute to the problematic situation, accept that information. Do not judge yourself. Try active listening with yourself. This information is added to the knowledge gained from active listening.

The importance of congruence

Part of this process is the concept of congruence. Be sure in all awkward interactions that all aspects of you are transmitting the same message. This includes your tone of voice, words, behavior, and body language. If your words say everything is great while your behavior or body language says something is different, people will not trust you.

4. Active Listening

Active listening is the act of putting yourself aside and listening to the other person. Many people are "difficult people" because they feel they are not being heard. After putting yourself on the map, listening may be all that is needed to listen to a resolution to the problematic situation.

In active listening, you must question your assumptions, your motivations, and your feelings to try to see the whole truth of the situation. Active listening is challenging to master, but it is a valuable life skill.

Active listening techniques include statements like: "To you, it must be like…," "you sound like… "sounds like you mean…" "I imagine that you are angry with this…" or "If you see the situation as… then I must seem like a real ass." No questions are allowed in active listening (questions put the other person on the defensive). If the person realizes you are listening, they will start to volunteer additional information. The five skills are:

- Paraphrasing
- Reflecting feelings
- Reflecting meanings
- Summarizing/Synthesizing
- Imagining out loud

Active listening works because it generates thought for both people. Use it before you act, argue, or criticize. Don't use it if you can't accept or you don't trust the other person. Also, don't use it to hide your feelings, or when you feel pressured.

(Partially adapted from Robert Bolton People Skills, Englewood Cliffs, NJ, 1979.)

5. Decide on a Course of Action

You've made some space, PYOM, and actively listened. By this time, you've probably determined your truth about the stressful situation. You now have to decide on a course of action. There are hundreds of coping strategies. We've listed some specific plans in this section.

One of the definitions of maturity is the ability to deal with difficult people/situations in flexible ways. In other words, you were being able to model your approach to the case rather than requiring the case to be handled in only one possible way. This flexibility allows you to be involved with a broader range of people and situations successfully.

There are general rules in deciding a course of action.

- The first question to ask: is this problematic situation yours to be upset about? Assume your boss storms in, ranting and raving about how the stupid idiots cut the maintenance budget on the 14th floor. This might be a tough situation for you, yet it's not your problem. There is nothing you can do to change that situation except adjusting to life within it. The next course of action is to either breathe to calm yourself or PYOM by giving your boss information about how you are affected by his anger. Always ask yourself if this situation belongs to you. People might try to dump their excess emotional junk onto you because you're a supervisor. Don't take it.
- Do not assume you are at fault. If you are at fault, acknowledge the facts of the situation. If not, do not join the person in berating yourself.
- Throughout the process of active listening, invoke your creativity. Listen to angles to resolve the problematic situation while letting both people save face.
- Focus on the specifics of the problematic situation. Don't dredge up the long history unless it is relevant. (Tom, this is the ninth time you've called about the heat.) Imagine there is some piece of information you don't have. You might find that his child is chronically ill. You can then solve the problem with a space heater in the child's room.
- Own the problem if it's appropriate. (We apologize for having to come back so soon on the same repair. We missed the outside bearing in our last repair.) Did you ever try to return a broken Craftsman hand tool at Sears? They have made a corporate decision to take responsibility for all failures of their hand tools. This makes it easy for their salespeople to deal with angry customers (Sorry Ms. Smith that the screwdriver broke, I'll replace it immediately, thank you for bringing it to my attention). We encourage all maintenance departments to view their jobs as professional service providers.
- Be willing to negotiate. Look at the options. The most frustrating experience for your subordinates or your customers is total inflexibility. Be prepared to work with people, and the frequency of stressful situations will drop.

- Whatever you do, don't paint yourself into a corner unless you want to be there (e.g., don't say to your boss that you'll be done with the repair when you're finished, and if that's not good enough he can fire you.)

(Partially adapted from Employee Productivity Consultants, *Dealing with Difficult People in the Workplace*.)

9 Be Alert for Unconscious Bias, Everyone Has It

Bias is your mind using shortcuts. Many biases are quicker to apply than thinking a problem through. They also seem to make your life easier. The problem with bias is that you are not seeing and not reacting to reality. So, while using bias is quicker, it also runs the risk of danger or massive misunderstandings.

TEN TRAPS OF THE HUMAN MIND

The human mind is powerful. Some naturally occurring biases mitigate power. These biases blind the person to, literally, what is right in front of their eyes. None of us are immune.

The alert supervisor is listening for which bias is in play and conduct their conversations and make their decisions with thought.

1. Fundamental Attribution Error

The attribution of a problem to individuals in a system rather than to the system in which they find themselves is so pervasive that psychologists call it the "fundamental attribution error."

> MIT System Dynamics Professor John Sterman and Nelson Repenning, published in the California Management Review in 2001 ("Nobody Ever Gets Credit for Fixing Problems That Never Happened")

The people are not the problem; the system is. We need to fix the system so the people can do better work. We need everyone's help to identify and fix the system, and only by working cross-functionally is it possible to succeed.

2. Normalization of Deviance

Normalization of deviance when people repeatedly accept a lower standard of performance until that lower standard becomes the "norm."

Colonel Mike Mullane NASA Astronaut relates the story of the Space Shuttle Challenger.

The NASA team accepted a lower standard of performance on the solid rocket booster O-rings until that lower standard became the "norm." They had become so comfortable with seeing occasional O-ring damage and getting away with it, the original standard, in which ANY O-ring damage was intolerable deviance, was marginalized.

3. Correlation Implies Causation

If two facts are correlated, are they related by cause and effect? A recent medical journal stated the results of an extensive study that kids who take one or more tablets of Tylenol have twice the amount of asthma. The news article leads with "Tylenol causes Asthma?" They wanted an unsophisticated public to think that "Correlation implies causation."

But in fact, correlation does not imply causation. The relationship between two variables does not automatically mean that one causes the other. In this case, kids who take Tylenol might have more colds and take more Tylenol. There is proper research that viruses are causally related to asthma. Correlation but no proof of causality is a prevalent problem in research.

4. After the Fact Therefore Because of the Fact

This mistake is so popular and well known that it has a Latin phrase describing it. This goes by the Latin "*post hoc ergo propter hoc*" and requires that one event occurs before the other. The phrase is Latin for "after this, therefore because of this."

It is a fallacy which states, "Since event X followed Y, then Y must have been caused by X." If I change grease in a bearing and the bearing fails, we might be clever and say that the new grease caused the bearing failure. But in fact, we need evidence beyond the mere time relationship for proof.

> It is a legitimate question (and a valuable one) to ask what happened before the failure or what has changed. The leap (without good evidence) from something happening first to that thing being the cause is Post hoc ergo propter hoc thinking.

5. Cherry-Picking

An observer who only sees a selected data set may thus wrongly conclude that most, or even all, of the data are like that. Cherry-picking can also is part of other logical fallacies. For example, the "fallacy of anecdotal evidence" tends to overlook large amounts of data in favor of some other cause.

Cherry-picking is the bane of medicine. Advertisements tout a new cure to high blood pressure or high blood sugar. If you track down the testimonials, you might find that it did work for that one person in that specific situation. The problem is that the burden of proof is much higher for a medication used by the public. Even with the higher burden of proof, the regulators sometimes get it wrong.

6. Name-Calling

Name-calling is a technique to use emotional arguments to substitute for rational arguments. People use the name-calling technique to incite fears (or arouse positive prejudices) with the intent that invoked fear (or trust). Both anxiety and faith are baseless. When this tactic is used instead of a logical argument based on evidence or experience, name-calling is thus a substitute for rational, fact-based arguments.

Current politics uses technique widely. There are some groups that (as a group) scare people (such as in the US immigrants). Name-calling invokes those feared groups to convince the audience that the other candidate will destroy our way of life.

7. The Fallacy of the Single Cause

"What was *the* cause of this?" Such language implies that there is one cause when instead, there were probably many causes. Even the phrase "Root Cause Analysis" encourages the idea that there is one cause that is the root of all evil.

Of course, we are looking for the cause that gives us the most leverage (least effort and most significant impact). This is not **the** Root Cause but the cause we can eliminate or mitigate with the least effort.

8. Regression Fallacy

If we assume some attribute is randomly distributed in any population (like free throw percentages), then the appearance of outliers (making all the shots or missing all the shots) would generally be followed by a regression to the mean for that population. An example is sometimes called the Sports Illustrated cover Jinx. An athlete has a great year. It is better than his past performance. He gets selected for a cover of Sports Illustrated. His subsequent performance regresses toward his old performance level.

This regression fallacy follows for all kinds of results in many fields. The logical flaw is to make predictions that expect exceptional results to continue as if they were average; People are most likely to act when the variance is at its peak (such as buying a stock). Then after results become more regular, they believe that their action was the cause of the change when, in fact, it was the normal regression reasserting itself.

The logic of the regression fallacy might look like:
The problem: The student did exceptionally poorly last semester.
The cause: We assume it is because he is not motivated
The action taken: We take away some privileges and punish him.
Result: He did much better this semester.
Punishment is effective in improving students' grades.

9. Circular Cause

The circular cause is where the consequence of the phenomenon is claimed to be its cause.

There are many real-world examples of circular cause and effect (many of them constituting virtuous or vicious cycles). Where the circular cause is a cycle, it is a complex of events that reinforces itself. A *virtuous* circle has favorable results, and a *vicious* circle has detrimental consequences.

- More jobs cause more money in people's pockets, which increases consumption, which requires more production, and thus more jobs.
- The expectation of an economic downturn causes people to cut back and spend less, which reduces demand, which results in layoffs, which means people have less money to spend, causing an economic downturn.

10. Third-Cause Fallacy or Sometimes Called Joint Effect

In this fallacy, there is a third, invisible factor driving both effects. The famous example is that a city's ice cream sales are highest when the rate of drownings in city swimming pools is highest. To conclude that one was a cause of the other is spurious. In this case, the invisible third factor could be a heatwave that was driving the ice cream sales and the increased pool use (and drowning)

WHEN YOU RUN MEETINGS, BE ALERT FOR COGNITIVE LAZINESS!

There are psychological tendencies that some people exhibit. These distortions are sometimes related to logical fallacies and sometimes create their unique problems.

Clustering Illusion

The fallacy is related to the clustering illusion, which refers to the tendency in human cognition to see patterns in random clusters where none exists.

Red Herring

A "red herring" is a tactic that seeks to divert the attention of an opponent or a listener by introducing a new unrelated topic. The best red herrings are ones that people have strong feelings about, so the saying of it will cause a reaction.

Rationalization

Rationalization (commonly known as making excuses) is the process of constructing a logical justification for an action, belief, or decision. It is a defense mechanism in which perceived controversial behaviors (mistakes of some kind) are explained rationally or logically to avoid the real explanation (the person messed up).

Texas Sharpshooter Fallacy

The Texas sharpshooter fallacy is a logical fallacy in which information fits into a pattern after the fact. The name comes from a joke about a Texan who fires at the side of a barn, then paints a target centered on the biggest cluster of hits and claims to be a sharpshooter.

Wrong Direction

The wrong direction is a logical fallacy where cause and effect are reversed. For instance, the statement: Alcohol consumption is high among people who have accidents. Therefore, accidents cause excessive alcohol consumption!

In other cases, it may only be unclear, which is the cause, and which is the effect. For example, Children that play violent video games are more violent than kids that don't. Video games make children more violent. This could easily be the other way around; that is, violent children like playing violent video games more than less violent children.

Jumping to conclusions: This is one of the common problems of the facilitation of RCA. The overwhelming tendency is to go directly from the issue to the solution skipping the essential step of developing causes

All-or-nothing thinking (splitting): Thinking of things in absolute terms like good-bad, right-wrong, or all, always, never. Few causes are so absolute.

Overgeneralization: Taking isolated cases and using them to make broad generalizations like all contractors will try to cheat you.

Mental filter: Focusing on particular and minor aspects of a situation such as the fact that the work order was not signed by the correct party to the exclusion of substantive issues.

Disqualifying the positive: Not seeing anything positive in a person or situation and shooting down any statement to the contrary. Overall negative attitude characterized by statements like "that will never work here."

Magnification and minimization: Magnification is typical in discussions about risk. People distort aspects of the situation through magnifying or minimizing them such that they no longer correspond to what happened.

Catastrophizing: Focusing on the worst possible outcome, however unlikely.

Emotional reasoning: Making decisions based on personal *feelings* rather than objective evidence. This tendency is regarded as a positive attribute in some circles. It could be positive if the contributor has a gut feeling leading a new direction, not apparent to others. If this new direction is then subjected to the same rigor as other causal paths, then it works for the analysis.

Should statements: People who are concerned more with what should be rather then what is going on. Also, people who have rigid rules that always apply no matter what is going on. Should statements could be related to wishful thinking.

Labeling: Labeling is very useful in categorizing facts, events, and situations. The downside is the tendency to label something and forget to look at the reality of the thing. This is also confusing the label with the actuality.

Mislabeling: It involves describing an event with language that is highly colored and emotionally loaded like John is a crook, John is lazy, John is X.

Personalization: In personalization, a person attributes causes to people who have little or no control in the situation. The giant downside is that this pattern is also applied to others in the attribution of blame.

Section II

Maintenance Management for Supervisors

10 Understanding and Avoiding Breakdowns

The better we understand the enemy, the more likely we will be successful in the battle. When most people think about maintenance, they think of the repair of breakdowns. While in this book, the focus is on avoiding failures or avoiding the consequences of failures; being good at repair is essential.

LEGACY

One of the legacies we fight is the old concept of the grease monkey mechanic. Through the PM effort and other approaches, we need to increase our reputation for professionalism. In other repair fields such as computer repair and copy machine repair, professionalism is a job requirement.

However, PM systems fail because PAST SINS wreak havoc on any supervisor trying to change from a fire fighting operation to a PM operation. Even after running for a few months, there may still be so many emergencies that you can't seem to make headway.

You face unfunded maintenance liabilities. The only way through this jungle is to pay the piper, repair, modernize, or rebuild yourself out of the woods. The investment is essential to change the story. Any sale of a PM system to top management must include a nonmaintenance budget line item for past sins.

Remember, wealth was removed from the equipment without maintenance funds being added to keep it in top operating condition.

BREAKDOWN STRATEGY (CALLED RUN TO FAILURE STRATEGY)

It's a common enough story that can make grown men and women cry. A truck breaks down, and the load is ruined. I saw a load of lettuce coming east from California after 10 hours in the desert sun (after the reefer quit). Pretty disgusting. Even the rabbits weren't interested!

That's half of the story for customers. Another truck had a load of colorant for plastic (this is a plastic resin containing concentrated color. The dye is mixed with the plastic pellets for injection molding of colored products). This truck had a minor failure that resulted in a roadside breakdown. Unfortunately, the driver was in a dead cell area. He used CB relays to get word to a repair facility.

His JIT (Just-in-time, where you must deliver a component of a product at a specific time) window closed, and the missing shipment shut the manufacturing process down. The problem for the shipper was that it was the first shipment of a $40 million contract for a new customer in a new industry. Ouch!

Breakdowns are not only not okay, but there are costs way above the repair costs. In fact, in some industries, the "other" costs are 10–100 times greater than the repair costs.

The costs of a breakdown look like an iceberg. Above the waterline are the repair costs, including parts, labor, and outside vendors. Below the waterline (90% of the iceberg) is the "other" costs.

LET'S GET CLEAR ABOUT WHO THE ENEMY IS

Our enemies are breakdowns and deterioration (not operations, purchasing, or engineering) and their consequences.

The enemy is any disruption to the value stream provided by the asset. One of the significant disruptions is breakdowns. We are trying to reduce failures, and we are also trying to figure out and minimize the consequences of failures that do happen. And deterioration, of course, can introduce other problems such as:

Maintenance costs
Production costs
Other costs

It seems like breakdowns don't occur at the most opportune moment. They seem to accumulate right before you have to leave to make a promised family evert. The most expensive and dangerous form of maintenance work: "Get the plant running" Not: "Fix the *real problem*."

Is downtime getting your department down? Evaluate the sources and impacts of downtime exposure. In most plants, the cost of downtime far exceeds the cost of maintenance. Strategies for predicting and minimizing unplanned outages

Losses Due to Downtime Can Dwarf Costs of Maintenance

Some equipment tries to communicate with us. What happens when we ignore the signs?

- What benefit is it to repair a chemical transfer pump before failure?
- What benefit is it to repair a heat exchanger before failure?
- What benefit is it to repair a wheel bearing before failure?
- What benefit is it to repair a drill press before failure?

When you act now, you spend money and labor now. But when you defer action, you stick the future maintenance department with an increasingly expensive problem

Delay and deferral create a problem: PAST SINS. PAST SINS are unfunded (delayed) maintenance work. This deferred work is called **unfunded maintenance liabilities**

Today's breakdown has a tail into the past, just like tomorrow's breakdown's tail reaches back to today. For some failures, your action or inaction today creates that future.

THERE IS A GOOD DEAL OF IGNORANCE ABOUT BREAKDOWNS

Some of this ignorance will affect your job, sometimes daily. If the breakdown is severe, then your managers might want some action immediately to make sure it won't happen again. This is a trap!

It is a trap because the causes of most breakdowns are outside the control of maintenance and require a multidisciplinary approach. The solutions must, of necessity, include production, other asset users, and anyone who touches the machine, like janitors, maintainers, or calibrators.

According to well-verified research, most of your breakdowns are random and have to do mostly with carelessness and have nothing to do with wear and tear of the asset or machine. That means when you use the tools of maintenance (like PM, inspections, etc.), you can impact only a small percentage of the breakdowns.

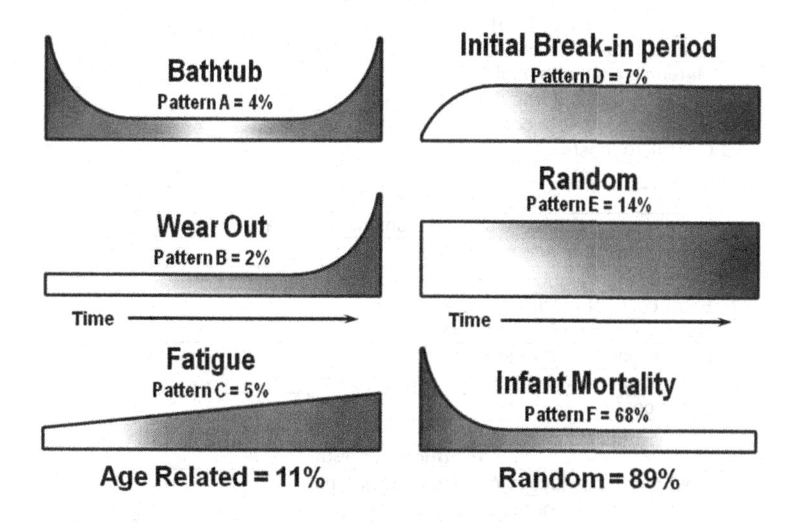

FIGURE 10.1 Work of Noland and Heap on the 6 failure curves. Shows random failure curves are more prevalent.

These curves, first published by Noland and Heap in 1978, were the result of their pioneering study for United Airlines. The left-side curves are age-related and are detectable in advance by inspection.

The right-side curves are random. In this case, random means they are equally likely at any time in the life cycle. They cannot generally be detected by inspection while they are unfolding.

The field Nolan and Heap created is called RCM (Reliability Centered Maintenance), and it is the most rigorous format of maintenance strategy. We will delve into RCM in more depth in the chapter called Maintenance Strategy Development.

Random Failures

Examples include reversing a machine before it stops or running a machine after a storm has contaminated the lubricant. For random failures, most root causes are due to carelessness or another outside source.

Breakdowns are commonly caused by someone's carelessness, contaminated raw materials, and bad parts. So, if you want to reduce failures, you must address carelessness. This includes carelessness from everyone who touches or impacts the asset.

Deep Dive into Breakdown Consequences

What are the costs and consequences of a major breakdown?

Breakdown Consequences
Maintenance costs
Labor with burden, overhead
Spare parts with the burden of the storeroom
Materials and consumables
Rental equipment and demurrage
Specialized vendor costs
Contractor costs
Additional maintenance costs
Extra costs due to core damage (destruction of a rebuildable part)
Extra damage of associated parts and the labor to repair them
Incoming airfreight instead of ground freight
Extra costs of outside parts and labor
Extra travel time for the mechanic
Extra repair time due to field conditions
Loss of morale among maintenance workers
Other costs
Production costs
Cost of lost production or price of downtime
Cost of airfreight of goods to the customer
Spoilage, contamination, or another compromise to the product
Loss of goodwill, loss of customers
Loss of sales from not being considered a preferred vendor.
Operator (or operations crew) idle time
Loss of smooth function and time to smooth production
Loss of status within the organization (political)
Loss of morale among production workers
Other costs
Other consequences
Safety, injury, sickness
Indirect safety costs, workman's comp, investigation
Environmental costs
Indirect environmental costs, fines

FIGURE 10.2 Table showing all the potential costs of a breakdown.

11 Practical Safety for Supervisors

Maintenance and repair activity can be dangerous. One of the supervisor's primary responsibilities is to keep their people safe by requiring safe work habits. This responsibility involves vigilance, skill, and knowledge of hazards. Safety requires understanding hazards, being safe, talking about safety, and interventions to correct safety issues.

The Hazards table

The table below is a reference hazards table that contains a list of all known hazards at most sites. When managing risk, there are four options:

- Accept it and do nothing.
- Remove or eliminate the risk.
- Mitigate the risk by reducing the severity or consequences.
- Transfer the risk, such as buying insurance, vendor contract, etc.

The hazards table also provides examples for eliminating the risk and mitigating the risk for each hazard.

When any of these hazards come into play, you might have an accident. An accident is rarely genuinely accidental, random event, but rather the result of a cascade of events or causes that end up in damage or injury.

WHAT CAUSES ACCIDENTS?

The dictionary definition of an accident is: "An unfortunate incident that happens unexpectedly and unintentionally, typically resulting in damage or injury."

Safety Risk	How to Eliminate It	How to Mitigate It
Airborne contaminants	Eliminate contaminants	Respirators
Falls from heights	Do work at ground level	Guardrails, fall protection
Falling objects	Don't allow work below	Hard hats, temporary roofs, shelters, better procedures
Eye damage from particles, chemicals, flash	Find another way to work, isolate this work, use robots if feasible	Safety glasses, face shields
Entrapment and crushing	Don't allow work where people can get caught or crushed	Improved procedures, safety shoes, proper clothing, improve chocking and blocking
Slipping and tripping		Housekeeping, proper shoes, equipment laydown areas well organized, high visibility paint and tape, good lighting
Chemical ingestion, skin exposure, breathing	Eliminate hazardous chemicals	Proper personal protective equipment (PPE), better procedures, access and understanding safety data sheets (SDS)
Radioactive exposure	Eliminate radioactive substances	Clear areas and paths of gamma rays, protective clothing, lead barriers, wear dosimeters and limit exposure
Fire	Eliminate flammables	Procedures, watch people, pick less-flammable chemicals
Electrocution	Lockout, do not allow work on or near energized circuits	Safe procedures, training, lockout
Explosion	Eliminate one or more of the components of the explosive mixture (there will always be a source of ignition)	Safe procedures, safety blanket gas, training
Asphyxiation	Do not allow entry into any dangerous areas	Gas monitors, gas alarms, air packs
Temperature stress	Do not schedule work during cold or hot times	Issue appropriate clothes, work time and temperature tables, cold vests, adequate cold water, shade, evening work

FIGURE 11.1 Hazards table with techniques to mitigate or eliminate the hazard.

Examples of items that mitigate hazards (left to right): Material and safety data sheets (now SDS) and books on how to handle hazardous materials and what to do if exposed to chemicals, lockout station with locks and tags to reduce the chance people will forget and eye wash station to reduce potential eye damage

FIGURE 11.2 Safety workstation includes hazard communications documents, locks and tags and eye wash station.

Why Did These Injuries Occur?

Generally, accidents are grouped into five categories. The areas are spread among many parts of the business. In one study by a global chemical company, the causes were:

• Specification	44%
• Changes after commissioning	20%
• Operations and maintenance	15%
• Design and implementation	15%
• Installation and commissioning	6%

It is essential to have a safe culture nurtured from the top of the organization. Environmental, safety, and health thrive in environments where management requires that they are reviewed at every step from the initial design to the plant process and procedures, even to the hiring of qualified operators and maintenance personnel.

I want to focus on the reliability of the equipment. Problems in all the categories above are causes of breakdowns. For example, in operations and maintenance, culture of expediency can contribute to procedures not followed, design flaws, and workers not being careful. With all that going on, it is easy to see that nonstandard situations can also occur. The result is the sum of everything, including all the decisions made, attitudes, and history.

Some of the accidents are the result of unsafe acts (hot work on a tank with an explosive mixture inside), failing to follow procedures (pressure testing with personnel in harm's way). Of course, most accidents like these have several causes at the same time.

We could take another cut at incidents to get to see more causes. The 5-Why technique features asking a simple question, "Why did this happen?" You keep asking why five times. That will get you deep into the roots of the problem. In this case, one question I would ask is, "Why was the person in harm's way in the first place?" The beauty of this technique is that answering "Why" takes you deeper into the incident. It is simple enough that almost anyone can do it.

ACCIDENTS AND QUALITY

This section is a parallel conversation with quality. It turns out that eliminating the causes of accidents also addresses many of the causes of mistakes and quality problems.

There are many common causes of accidents in the workplace. Some accidents have overlapping reasons and accountabilities.

Here is a big list of some causes of accidents

Remember: The more you manage, the fewer causes are "out there."

Management

- Unrealistic expectations, pushing too hard
- Not enough money or time to do the job correctly (improvisation)
- The mentality to cut costs regardless of consequences
- Attitude to ignore the advice of maintenance, engineering and reliability professionals
- Poor planning
- Not demanding safety be discussed and dealt with at every stage of an activity
- Acceptance of temporary repairs with no plans to remediate
- Interruptions from managers

Processes and procedures

- No risk management
- No hazard identification
- Boilerplate LOTO (some add LOTOTO for Try Out)
- Ineffective hazard permitting (i.e., hot work)
- Lack of a permit to work system, when needed
- Inadequate PPE for the level of danger
- Different rules for management when they are in the shop
- Adherence to regulations optional

Supervision

- No instruction
- Inadequate instruction (i.e., didn't communicate)

- Incorrect direction
- Absent supervision
- Bad supervision
- Improper scope, no scope, wrong scope
- Inadequate communication between trades or shifts
- No wiring schematics
- The supervisor not standing for safety

Engineering

- No drawings
- Drawings wrong
- No as-built drawings
- No operations and maintenance (O&M) manual
- Equipment operated beyond design capacity
- A machine used for something it was not designed to do
- Bad design for use
- Designed with difficult access
- Poorly designed equipment piping, wiring, or foundation
- No testing, no commissioning
- Old equipment at the end of the life cycle with multiple unfolding failure modes

Operations

- No standard operating procedures (SOP)
- Lots of shortcuts and tribal knowledge needed to operate

Conditions of people who can contribute to accidents

- People untrained (i.e., ignorance)
- Trained people without experience (i.e., new graduates)
- Trained people without confidence
- Anger at the company (i.e., sabotage)
- Low morale, don't want to do the work
- Lousy attitude (rare by itself, usually accompanies another cause)
- People who don't have the capability (e.g., intelligence, strength, flexibility, endurance, visual and auditory acuity)
- People feeling frustrated and making mistakes, such as not being able to locate things
- People are drugged, legally or illegally
- People are drunk, hungover
- Things preoccupy people outside of work
- Things preoccupy people at work (e.g., personal conflict, layoff, merger, etc.)
- People are tired from long hours or moonlighting
- People dehydrated, low blood sugar
- People off their usual prescription medications or adding new drugs
- People currently sick or not wholly healed
- Injury not healed yet

FIGURE 11.3 Fireproof cabinet.

Tools

- Wrong tools
- Broken tools
- Cheap tools
- Inadequate capacity of tools
- Improvised tools
- No tools
- Don't know how to use tools available
- Lack of PPE

Materials (parts, disposables, consumables, free issue)

- No material, lack of enough material
- Wrong material, but right part numbers
- Wrong material, incorrect part numbers
- Slightly wrong material (i.e., make it fit or can be adapted to work)
- Cheap materials

Working conditions

- Bad lighting, such as too dark, wrong color for work
- Need for magnification
- Slippery
- Bad air, smells, chemicals
- Dusty
- Bad or inadequate work platforms
- Too hot or too cold
- High humidity
- Full sun

- Rain, snow, sand, or dust storm
- Lightning, storms
- Graveyard shift
- Working at heights with fear of heights
- Other environmental factors

Note some of these causes can be addressed by supervision, and other items require investments from management. In all cases, the supervisor is the representative for the safety of their crew.

JOB SAFETY ANALYSIS

Risk management (and risk identification) is best done as a team. The reason is that people from different backgrounds will see various potential hazards. Imagine the input of Millwrights, Riggers, Operators, Engineers as well as safety people to the risk equation. Of course, we don't want to go overboard because if we do, there will be a waste, and it will not be a lean process.

The planner breaks the job down into steps. This process helps identify resources and makes it easier to estimate. One other process made more manageable by work breakdown is the identification of hazards. As an example, let's look at a job to remove and replace a large horizontally mounted pump. Further, let's agree that the job steps are:

Job Plan Remove & Replace Pump	
Step number	**Activity**
1	Permit, lockout, tag-out
2	Shore up discharge section
3	Drain and blind
4	Un bolt, rig and remove spool on the suction side (171#/foot in 24", 3' length = 513 pounds)
5	Rig and Remove pump with crane (2350 pounds)
6	Clean base scrape flat
7	Replace pump, bolt down, align
8	Remove blinds
9	Get a testing permit, release locks, tags
10	Test and benchmark system
11	Remove shoring and clean area
12	Return to operations and clear all permits

FIGURE 11.4 Job plan (steps) to remove and replace a pump.

Risk Management (adapted with permission from the author's *Managing Maintenance Shutdowns and Outages* Industrial Press). The three steps as parts of the planning process are risk identification, risk quantification, and risk response.

The first two identification and quantification are sometimes grouped under Risk Analysis or Risk Assessment.

JSA is the process we use to detect hazards and decide what to do with them. The purpose of a JSA is to ensure that the risk of each step is reduced to ALARP (As Low as Reasonably Practicable). We next look at each step and see if any of the hazards from the list are likely, probable, or possible (high, medium, or low probability).

If we take just a few steps from the job plan, we can see what risks are present, and based on the impact and probability of occurrence, we can decide on a course of action.

Job Plan to remove and replace a pump with Hazards and Mitigation			
Step	**Activity**	**Hazard**	**Steps and PPE to mitigate**
2	Shore up discharge section	Entrapment and crushing, Falling objects	Mitigations adequate design, hard hats, steel-toed boots
3	Drain and blind	Airborne contaminates, Asphyxiation, Chemical ingestion, skin exposure, breathing, Eye damage (particle, chemical, flash)	Fresh air, gloves, full-body moon suit (if needed), face mask,
4	Un bolt, rig and remove spool on the suction side (513#)	Entrapment and crushing, Falling objects	Steel-toed shoes, rigging standards, an inspection of straps and chokers, etc.,
5	Rig and Remove pump with crane (2350#)	Entrapment and crushing, Falling objects, Asphyxiation	Procedures to clear the lift path, formal lift plan, test air before getting too close, Steel-toed shoes, rigging standards, an inspection of straps and chokers, etc.,

FIGURE 11.5 Listing of hazards for the job plan by job step.

Safety is less expensive if it is planned into the job rather than tacked on afterward.

Why are people placed in a position to get hurt? The answer to that question is straightforward and might not occur to maintenance professionals. This reason is at the core of a high percentage of accidents. If we look at more maintenance-related fatal incidents, we can start to see a pattern.

- *One described a tragic double fatality of welders in a petrochemical plant. An argon cylinder with a defective regulator was leaking into the vessel. The first welder went in and collapsed; the second, went in after him, collapsed, and they both died.*

Think of it, when we fix some of the root causes of the fatalities or injuries, whether they are caused by behavior, culture, design, process, or procedure, more workers will go home at the to their families whole and intact. There are many reasons for these injuries and fatalities. Some of the common ones include traffic accidents, falls, and a whole host of injuries and, unfortunately, fatalities from maintenance work.

I want to discuss maintenance-oriented injuries and fatalities. Some examples of severe accidents from OSHA records:

- *A massive explosion destroyed a large storage tank containing a mixture of sulfuric acid and flammable hydrocarbons at the Motiva Enterprises Delaware City Refinery. Nine contract workers were injured, one fatally. Sulfuric acid from collapsed and damage tanks polluted the Delaware River. The explosion occurred during welding operations to repair a catwalk above the sulfuric acid tank when welding sparks ignited flammable hydrocarbon vapor.*
- *Crews were doing maintenance work on a generator. During a pressure test, a manway blew off, striking two contract workers injuring them critically.*

Why are people placed in a position to get hurt? **The related question is, how are reliability and EHS (Environmental, Health, and Safety) related?**

Reason 1: Something broke! The breakdown caused the maintenance person to go into harm's way. So, a lack of reliability causes death and injuries.
Reliable equipment removes this cause—one of the common causes of accidents. We could be even more specific. A machine running as designed does not require people to enter a confined space, repair (and touch) exposed electrical wires, pressure test a generator, sitting on top of a tank and welding, or even falling off of a ladder.

How is reliability related to safety? Reliability removes the risk from the equation, and the worker is not in harm's way. If no one were welding above the tank, then the explosion would not have happened; if there were no repair needed, no one would be up on the ladder or the roof.

- Something breaks down and has to be repaired.
- The breakdown causes a worker to be in harm's way.
- Reliable equipment does not require maintenance workers to be put into harm's way.
- The best solution to a hazard is to eliminate it.

Reason 2: Size and scope of repair is smaller (due to PM) making for safer repairs.
ExxonMobil has reported the second part of the equation. They studied their maintenance-related accidents and found: "Accidents are five times more likely while working on breakdowns then they are while working on planned and scheduled corrective jobs."

High reliability implies an effective PM program that catches deterioration before it causes a failure. The asset is not yet broken. Any work to be done will be smaller and have reduced scope.

- PM activity catches deterioration early in the process before failure.
- At that point, the repair is smaller, safer, and more manageable. Fewer EHS (Environmental, Health and Safety) incidents
- It also gives managers more time to plan and deal with hazards.

Reason 3: Hazards are eliminated or mitigated in the planning process.
High reliability also implies that the maintenance planners have time to plan the job correctly. One aspect of planning is to consider all the hazards and figure out and describe a way to accomplish the work safely. The job plan that an experienced planner develops will reflect the safe way to do the job.

A planner should look at every job and see if any common hazards are present. Hazards would include airborne contaminates, fall from heights, slipping and tripping, falling objects, eye damage (particle, chemical, and flash), Chemical (ingestion, skin exposure, breathing), asphyxiation, radioactive exposure, fire, explosion, electrocution entrapment, and crushing temperature stress.

Every hazard identified is then eliminated (best option) or mitigated (second-best option). The safest plants are the ones where the safety of the workers is considered at every step in the job preparation process.

- The planner plans the job to minimize downtime
- The planner is specially trained to look for hazards to safety, health, and the environment.
- Planners will mitigate or eliminate the hazard in the plan before the crew even leaves the shop.
- Result in fewer EHS incidents, more reliable equipment

Reason 4: Planned jobs allow fewer opportunities for the maintenance worker to improvise.
Improvisation is statistically less safe than following the job plan with the correct tools and spares. One of the building blocks of a reliable culture is adequate maintenance planning. Without planning, the workers are forced to make do with what spares and tools they can find. To do their job, they may have to improvise to make things work. Improvisation might be great in the theater but can be deadly in maintenance. I guess that the following worker was making do with improvised support:

The worker was performing maintenance on the back of a trash truck. He places a 4X4 to support the ram. The support slipped and fell out of the way, and the ram came down on the worker.

- Improvisation is great in comedy and deadly in maintenance.
- Adequate time for job planning means having the right tools, spares, equipment, skills, and drawings when the job starts.
- Results in fewer EHS incidents and better reliability

Action Items for Quality and Safety

Management action items to transform the culture require minor modifications to the weekly and monthly KPI (Key Performance Indicators) used to run the plant or facility and for bonuses.

1. The ratio of emergent maintenance work to planned and scheduled maintenance work should be above 80% planned and scheduled.
2. PM performance above 95%. More than 95% of the PMs generated are completed in ±10% of the PM interval (30-day PMs are done between 27 and 33 days).
3. Schedule compliance above 85%. That means more than 85% of the jobs scheduled are completed sometime the week they are scheduled.
4. MTBF for significant assets on an improving trend.

What if one of your plants doesn't measure up? Then it is time to talk through the problem, study your best plants, and consult experts. However, changing a culture takes time and will take three attributes:

- Follow-through to keep people's eye on the goals
- Resilience to get the plant back on track when the program goes off the track
- Positive attitude—Just like teaching a child to ride a bike, keep up a positive, encouraging attitude. Don't punish honest mistakes; make sure your people learn from them.

You can find checklists for safety inspections and audits in the Appendix

Hazard Communication

SDS (Safety Data Sheets) basics

What every maintenance supervisor needs to know?
The most common mistake that organizations make with the SDS and the chemicals is that the inventory of chemicals in use does not correspond to the SDS sheets in the notebook or file. The fine, as in other areas, run from $7000 per incident per day to $25,000 per episode per day.

1. Consider computerizing the whole process. OSHA has accepted the presence of the SDS sheets on the computer. With networks, you can avoid trouble by having a company-wide file available to everyone.
2. It is the manufacturer of the chemical that is responsible for getting the SDS to you. You are liable to notice if you have it or not.
3. A significant source of inspections originates from disgruntled employees anonymously calling OSHA.

SDS highlights:
The exact format of the sheet varies by manufacturer, but there are general sections that contain critical information that every employee should know. There are as many as 15 parts to the SDS sheet.

<table>
<tr><td>Part #1</td><td>OSHA hazard</td><td>This section is an overview of the hazard. Read it carefully to determine if a significant risk exists.</td></tr>
<tr><td>Part #3</td><td>Precautionary label information as prescribed by the EPA</td><td rowspan="2">These sections go through the specific first aid steps for all types of exposure, including ingestion, inhalation, skin contact, etc. If you work with dangerous materials, study these sections, and be prepared.</td></tr>
<tr><td>Part #4</td><td>First aid</td></tr>
<tr><td>Part #6</td><td>Toxicological Information</td><td rowspan="2">While this section is designed for doctors and scientists, a quick read through these sections would tip you off if the material is very toxic or very dangerous to the environment.</td></tr>
<tr><td>Part #7</td><td>Environmental Toxicology</td></tr>
<tr><td>Part #9</td><td>Fire</td><td rowspan="2">This part will alert you to a potential fire hazard or a potentially deadly reaction with another chemical. In some cases, mixing two safe chemicals results in an explosive reaction. For example, the gas given off by mixing bleach and ammonia is deadly.</td></tr>
<tr><td>Part #10</td><td>Reactivity</td></tr>
<tr><td>Other parts</td><td>Transit, handling, spill procedures, regulation</td><td>All of the other information necessary to use, transport, clean-up chemical</td></tr>
</table>

FIGURE 11.6 Sections with descriptions of the SDS for a chemical.

12 World-Class Maintenance

World-class maintenance is an aspiration. It is not a thing; it is a goal. Like all human virtues, no one is completely happy, generous, or always compassionate. We aspire to these things. World-class maintenance or its bigger brother world-class reliability is something we want to be moving toward, knowing once we get there, it will have moved onward.

ATTRIBUTES OF WORLD-CLASS MAINTENANCE DEPARTMENTS

- **Basic**
  - Do no harm to employees, communities, or the environment.
  - Actively manage risks.
  - Manage assets if they continue to provide value.
  - Waste nothing.
- **A new way to view maintenance activity**
  - Maintenance is a part of asset management.
  - Asset management is managing assets from Lust to Dust.
  - When you do work on or use an asset use precision technique.
  - Reduce all cost inputs—the less resource used, the better.
  - The ongoing need is to improve output, yield, and quality.
  - All activities are accurately accounted for in work orders with enough details.
- **New ways to view problems**
  - Solve your problems permanently.
  - Willingness to run controlled experiments.
  - Spend 1% of your effort improving and not just fixing!
- **Bad News** **Good News** **Other news**
  No rules. No rules. You get to try out new rules.
- **Hard numbers are king**
  - Why? To measure continuous improvement.
  - Willingness to use sophisticated tools of statistics, finance, accounting in maintenance analysis.
  - Willing to try AI and advanced analytics in maintenance analysis.
  - Another way to say this: **Goodbye** seat of the pants, hello—analysis-driven maintenance decision-making.
- **Always: Focus on service to the customer or focus on adding value to the customer**
  - Encourage the customer to participate in the maintenance effort.
- **People are the key: Right people in the right jobs with the right competence**

- Teams have different points of view and can see defects and solutions from different angles.
- Fade traditional barriers to information.
- Continual and cross-training.
- Attachment to people rather than technology.
- Layoffs are always the last option.

The Goal: Organization's AIM (Mission and Vision) is fully supported.

The result of world-class maintenance is a powerfully self-motivated workforce, and excellent execution of maintenance, well-supported customers.

Essential Question: Is your company ready? If not, your organization will be eaten by one who is! W.E. Deming says: "You don't have to do this; survival is not compulsory."

ANOTHER LOOK AT WORLD-CLASS PERFORMANCE

Let's take another look into the world-class maintenance. So-called "World-Class Maintenance" is something you read about in your maintenance journals and hear about at conferences.

It is an amorphous concept and does not exist. It is not real, BUT:

- It can have real benefits.
- It creates an opening for action.
- It can cause people and their organizations to stretch themselves and develop.
- If someone claims they can get you there, it is **guaranteed** they will take you for a ride!
- Ultimately it is aspirational.

As an aspirational domain, some of the things you will find once you find it:

- Everyone a reliability leader
- Good risk management
- Defect elimination
- Precision everything
- Proper commissioning
- Stick to schedules
- Experimentation
- Time and money to work out bugs
- Design for reliability
- IIoT Sensors
- Big data
- Advanced analytics, AI
- Prescriptive Maintenance
- RCA and fix it permanently
- Maintenance prevention
- Poka-Yoke
- Maintenance is "at the table" for strategic discussions

By the way, most of the things mentioned above are cultural!

13 Asset Management

We want to manage the organization's assets to provide maximal value, the lowest risk for the most economical cost. Companies have managed their assets from the beginning. The recent challenge is that organizations are so big that consistency, ability to work together, or even knowing what the best decision for the whole company is challenging.

Asset management and ISO 55,00X are attempts to make the process consistent. They are also documents that align different parts of the organization to work toward the common good.

Uptime Elements: Introduction to the Uptime Elements asset management framework.

Asset Management: The ISO standard 55000 is coming. What it means and what it doesn't mean?

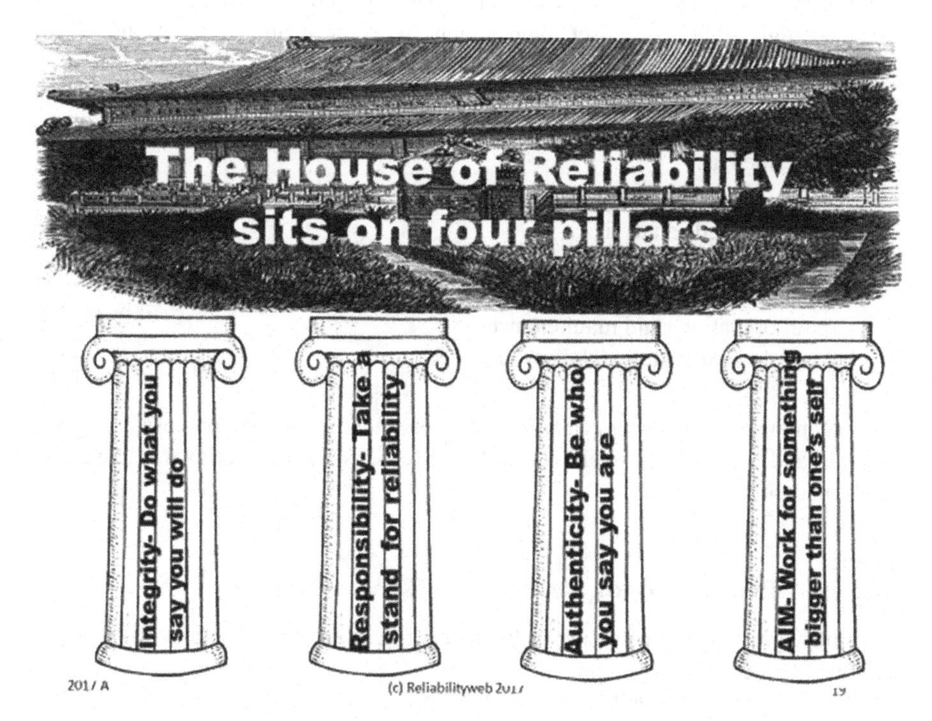

FIGURE 13.1 Reliability is supported by 4 pillars.

Uptime Elements – Uptime Elements designed by Terrence O'Hanlon at www.Reliabilityweb.com is a framework for asset management and reliability.

Reliability is a business imperative. Top management sets the aim of the organization, creates the definition of value, and sets the tone for asset performance.

Leadership is responsible for:

- Championing reliability
- Providing AIM (mission, vision, values)
- Staffing with enough people who have the appropriate competencies
- Setting a few key performance indicators
- Align stakeholders who can enable or disable reliability.

Information about the Uptime Elements® and the CRL® test are presented courtesy of www.Reliabilityweb.com Information can be found at www.Reliabilityweb.com and www.Maintenance.org

KNOWLEDGE DOMAINS OF THE UPTIME ELEMENTS

Asset Management (AM) (Yellow)

Initiated because the mission, vision, and values of the company need low costs, low risk, and high reliability. Asset management decision at high levels from SAMP (Strategic Asset Management Plan)

Asset management takes all phases of an asset's life-cycle into consideration when designing a managing system that delivers value. Typical asset lifecycle phases can be described as:

- Business needs analysis.
- Asset planning.
- Asset design and creation.
- Asset operations and maintenance.
- Asset decommission and disposal.

An asset management system ensures that an organization has the processes, support systems, and people to make effective whole life decisions that optimize value delivery from assets.

Reliability Engineering for Maintenance (REM) (Orange)

Reliability engineering with corporate engineering work on the details of the assets and operating conditions.

This is the only knowledge domain that can reduce failures.

Every task in a reliability program should target:

- Prevention of a failure cause
- Prediction of a potential failure cause
- Meeting a regulation or legal requirement

Reliability engineering for maintenance will:

- Inform, educate, and enlighten leadership and team members.
- Create a leadership line of sight from top management to plant floor regarding value, criticality, reliability, and risk.
- Direct your asset condition management and work execution management tasks and decisions.

Asset Condition Management (ACM) (Green)

Knowing, in real-time, the condition of your assets is essential. ACM is one of the significant data streams into the AI module.

It shares four attributes:

- Alignment of objectives
- Risk-based decision-making from a position of knowledge
- Long-term strategic view
- Transparent and consistent decision-making

There are three fundamental principles used in this domain.

- ACM reduces/eliminates defects from entering the organization through the application of precision lubrication techniques, precision alignment techniques, and precision balancing.
- ACM utilizes condition monitoring technologies and nondestructive testing technologies to provide early detection of possible failure modes to optimize planning, scheduling, and material requirements.
- It also includes a unified system for information management and decision support related to managing the condition or health of equipment and assets.

Work Execution Management (WEM) (Blue)

All work done is managed by WEM for safety, cost, and quality. This includes your backlog, work order, and all reporting.

- This domain guides organizations with the processes and tools they need to execute and manage work efficiently and effectively.
- As new maintenance tasks and MRO spare parts, requirements begin to flow from their Reliability Engineering for Maintenance (REM) domain.

Leadership for Maintenance (LEM) (Red)

Leadership is given its marching orders from asset management. The parameters are decided upon and carried out by leaders. Direction determines the scope of the effort with priorities and some specifics in consultation with other stakeholders.

ISO 55000

A set of three standards issued by the International Organization For Standardization

ISO 55000: Terms and definitions
ISO 55001: Requirements
ISO 55002: Guidance

It will be harmonized at a high level with other managing systems like ISO9001 (quality), ISO14001 (environmental), and ISO50000 (energy sustainability).

Published on January 2014, the standard DOES NOT tell you how-to perform asset management or how to deliver value and reduce risk—it excludes explicitly technical standards and guidance.

The decisions and processes will deliver the value in managing the risk opportunities around acquiring, operating, maintaining, and decommissioning your assets.

SYSTEM FOR ASSET MANAGEMENT AND RELIABILITY

- The strategy highlights the framework to manage and assure the outcomes and objectives.
- Policies, strategies, and plans provide answers to "Aligned Objectives" and "Transparent and consistent decision-making."
- Policies, strategies, and plans contribute to the execution of the "Long-Term Strategic view" and "Risk-Based decision-making."

Policy

- Describes the overall approach to asset management
- Its purpose is to establish the intentions and direction of asset management
- High-level objectives for asset management
- Guides high-level decisions
- The clarity for the "rules of the game" for decisions
- Consistency for asset management strategy, asset management objectives, and individual asset management plans

Asset Management: Find the Value and Minimize the Risks

Consider the list below. These are some values organizations have derived from active asset management with sensible reliability programs.

1. Reduce the size and scale of repairs	Risk
2. Reduce downtime	Value
3. Increase accountability for all cash spent	Risk
4. Reduce the number of repairs	Risk
5. Increase equipment's useful life	Value
6. Increase operator, maintenance mechanic and public safety	Risk
7. Increase the quality of output	Value
8. Reduce overtime for responding to emergency breakdown	Risk
9. Increase equipment availability	Value
10. Decrease potential exposure to liability	Risk
11. Reduce spare, stand-by units required	Value
12. Ensure all parts are used for authorized purposes	Risk
13. Increase control over parts, reduce inventory level	Risk
14. Decrease unit part cost	Value
15. Improve information available for equipment specification	Value
16. Lower maintenance costs through better use of labor and materials	Risk
17. Lower cost/unit (cost/ ton of coal, cost/part, cost? student)	Value
18. Improve the identification of problem areas	Risk
19. Special risks or values	?

FIGURE 13.2 Benefits from managing reliability.

New York City Metropolitan Transportation Authority Asset Management Policy

The Metropolitan Transportation Authority (MTA), with 50,000+ employees, provides over 5 million subway trips a day as well as 1.8 million bus trips. This policy statement sets the rules for all MTA divisions. Included in this policy are the track, stations, roads, and bridges to support the effort.

IDEAS ABOUT ASSET MANAGEMENT

- Asset management strategy and planning is an essential first step
- Align an organization horizontally across departments as well as vertically from top management to the shop floor
- This process itself has the potential to deliver value as stakeholders begin to "pull the rope" in the same direction

Asset Management Policy

Deliver safe and reliable services while focusing relentlessly on improving customer experience, responsibly managing our finances, and continually improving our capital assets

To all Metropolitan Transportation Authority (MTA) employees and contractors...

The MTA manages and operates one of the largest infrastructure systems in the world. We are duty bound to do this in a manner that is safe, robust, efficient, and sustainable. How we make and implement decisions about our extensive asset base must be rooted in good asset management.

With this in mind, we have established an Enterprise Management System. This will support our work every day and lay the foundations for continuously improving as we face more challenging scenarios from financial, technological, and urban planning perspectives.

This Asset Management Policy sets out the core principles of our Enterprise Management System. The core principles are applicable across the entire MTA Family and provide the criteria to guide us in every decision we make throughout the end-to-end lifecycle of our assets. By aligning to this policy we will enhance communications with our stakeholders with regards to the performance of our system and our investment allocation.

1. **Safety** is and will always be our number one priority: we will reduce safety risks for our customers, our employees, our suppliers, our neighbors, and everyone potentially impacted by our work.
2. The **Reliability** of our assets and the technologies we use will be optimized to deliver the system capacity, capabilities, and schedules we are expected to achieve.
3. We will provide safe, reliable, clean and accessible transportation services as well as timely and accurate service information to deliver the **Customer Experience** our customers deserve.
4. We will invest our capital and operational funding efficiently and effectively, ensuring that every dollar is spent based on criticality and risk to assure **Value for Money**.
5. All investments will consider the **Sustainability & Resilience** of our asset base to sustain a State of Good Repair, meet current and future service demands, and also consider the long-term societal and environmental impacts.
6. The women and men leading our **Organization** will be enabled to continuously improve their knowledge and skills, build their teams and the MTA, and reflect the wonderful diversity of our region.
7. We will ensure and be able to demonstrate the **Compliance** of everything we do with relevant regulation, legislation, standards and industry codes.

These principles are the basis for how I make decisions, how you make decisions and how we continue to keep New York moving.

Thomas F. Prendergast, Chairman and Chief Executive Officer

Version 0.5, January 3, 2017

FIGURE 13.3 Asset Management Policy for MTA (New York City subway, bus, light rail, and facilities).

14 Reliability Strategy Development

Reliability strategy development (RSD) is the process of deciding how and why to take care of each asset. RSD considers the risks and the value provided and provides the strategy. It is an essential first step to a prioritized and rationalized investment of talent and time. Scarce resources drive engineering decisions.

RELIABILITY STRATEGY DEVELOPMENT

Sound asset management practices manage two main attributes of an organization's assets. One is getting value from the asset. The value is how the asset contributes to the mission of the organization. The second is the management of risk.

Reliability strategy development explicitly looks at potential risks and predicted value derived from the asset to come up with a suitable strategy. One-size strategy does not fit all assets even the same asset in different services might require different approaches. We use RSD techniques to maximize the use of scarce maintenance and reliability engineering resources.

Criticality Analysis

The first step reliability strategy development is criticality analysis, which is concerned with evaluating all types of risks. We look at several classes of risk. Each has its consequences. This process focuses on risks that are more prevalent or more consequential.

- Hazards (safety)
- Health
- Environmental
- Downtime
- Breakdown cost
- Reputation and customer service
- Legal liability
- Fines
- Criminal

Risk management is the process of identifying the risks, quantifying the threat by the impact on the organization, and then evaluating the chances of the risk event happening. After the asset criticality evaluation, the organization must have strategies and action plans to manage the risk.

Impact: Each organization must define the impact

- Catastrophic
- Severe
- Moderate
- Minor

Probability: How often do we expect this risk?

- Low <10 years
- Moderate 3–9 years
- High, medium 1–3 years
- High <1 year

Risk Management Strategies

- *Eliminate the risk*: Can the risk be avoided by changing techniques (build something on the ground rather than in the air, get rid of the chemical, work de-energized, work cool, etc.) or reengineering.
- *Mitigate (or manage) the risk*: Mitigate to either reducing the probability of the risk happening (clear lift paths) or reducing the severity of the outcome (safety glasses).
- *Transfer the risk*: Reduce the financial impact of the risk (buying fire insurance) or outsource the manufacture some combination.
- *Accept the risk*: You decide that the threat won't happen (like a meteor crashing through the tower, so you must start over) or you accept the consequences.

RELIABILITY STRATEGY OPTIONS

The criticality of the asset determines the strategy. Reliability is a business strategy for a business risk (downtime, missed shipments). Reliability strategies hopefully eliminate the risk or at least to mitigate the consequences of the risk.

RCM: Reliability Centered Maintenance
FMEA: Failure Modes and Effects Analysis
PMO: PM Optimization
PM: Preventive Maintenance using common sense or CMMS history
PCR: Planned Component Replacement
RTF: Run to Failure

In addition to pure maintenance strategies, some common design strategies will help mitigate adverse consequences, including:

- Hot swapping: being able to change a component without impacting production.
- Build-in reservoirs of power or raw materials to allow the system to run without input.
- Redundancy: Being able to switch the load to a spare unit that is always available (also called in-line spare)

15 Advanced and Rigorous Maintenance Strategies

The supervisor might or might not be involved in the details of the design of the RCM or PMO strategy. It is vital to understand these processes since, after roll-out, you will be responsible for sustaining the program. Knowledge will also help you with ad hoc reviews of asset history and PM tasking.

FMEA, RCM, and PMO comprise the most vigorous, comprehensive, and advanced maintenance strategies. They are sophisticated, but they have been around for decades (PMO is the youngest). These strategies are used on whichever of your assets are critical to the mission of the organization or are particularly dangerous

Failure modes and effects analysis (FMEA) is a systematic, proactive method for evaluating a process to identify where and how it might fail and to assess the relative impact of different failures, to determine the parts of the process that are most in need of change (Institute for Healthcare Improvement).

FEMA is also a tool for identifying potential problems and their impact.

Begun in the 1940s by the US military, FMEA is a step-by-step approach for identifying all possible failures in a design, a manufacturing or assembly process, or a product or service.

- **"Failure modes"** means the ways, or modes, in which something might fail. Failures are any errors or defects, especially ones that affect the customer and can be potential or actual.
- **"Effects analysis"** refers to studying the consequences of those failures.

FMEA includes a review of the following:

- Steps in the process
- Failure modes (What could go wrong?)
- Failure causes (Why would the failure happen?)
- Failure effects (What would be the consequences of each failure?)

FEMA sequence

- What are the components and functions they provide?
- What can go wrong?
- What are the causes?
- What are the effects?
- How harmful are the effects?
- How often can they fail?

- How can this be prevented?
- Can this be detected?
- What can be done; what design, process, or procedural changes?

Process Step	Potential Failure Mode	Potential Failure effect	Severity	Is it detectable?	Risk Priority Number	Action recommendation	Responsible	Actions Taken and re-analysis

Initial Development of FMEA — Proposed Improvement — Post Improvement

FIGURE 15.1 FMEA worksheet.

Free guide: https://www.isixsigma.com/tools-templates/fmea/fmea-quick-guide/

RCM IS THE RIGHT CANDIDATE WHEN CRITICALITY IS HIGH

In 1978, there was a report written that transformed maintenance. It was called Reliability Centered Maintenance (now known as RCM) by Noland and Heap. They worked for United Airlines (the Assistant Secretary of Defense, sponsored the study).

This work was the result of the techniques used to design aircraft. The first jet that used it was the 747, which flew in 1969. The 747 was more reliable and needed less maintenance than the prior generation 707 and was quite a bit bigger and more complex.

RCM

- SAE JA 1011 and 1012.
- The engine of RCM is an FMEA analysis of the components of the asset.
- Understand that most failures are not linked to the asset age.
- Moving from efforts to predict life expectancies to trying to manage the process of failure.
- The design and development phases of the asset lifecycle are the best time to apply RCM.

RCM Philosophy

When maintenance focuses on reliability, your efforts attempt to eliminate the failures and, more importantly, the consequences of the failures. How much time a month do you personally spend thinking through failures to fix the root cause permanently so that they never darken your doorway again?

1. The consequence of failure is more important than the failure itself. Potential failure is treated based on what are the levels of losses (both financially and human). The same valve will be treated differently under RCM if, in one case, it is a critical component in an intensive care unit (hospital) versus a part of a swimming pool system.
2. If the consequence of the failure is death or significant environmental damage, then the risk has to either be designed out of the system or PM tasks designed so that the probability of an unanticipated failure is close to zero or detects it in time to repair.
3. For failures that cause downtime and repair costs to design a PM task that costs less to perform annually than the price of the failure plus the value of the downtime times the probability of the failure occurring that year. In other words, the cost of the task has to be cheaper than having a breakdown.

One of the most critical aspects of RCM is its holistic approach to equipment. This comprehensive approach to equipment comes at it from many different vantage points, including operations, maintenance, storage, energy sources, and others.

Breakthrough: Consequence-Driven Maintenance

They decided to look first at the consequence of a breakdown (broken into categories like deaths, environmental catastrophe, airplane downtime, and cost of repairs) to see if it was worth it to spend the time to reduce the chance of failure to as close to zero as possible.

The solution is a defense in depth.

Defense in depth is a concept that is used in mission-critical applications and has an excellent application to how you manage your effort.

RCM Also Focuses on Hidden Functions

The hidden failures are a big part of the functional analysis in RCM. Hidden failures occur in protective systems. Operators would, under normal conditions, will not notice the hidden system problem. The investigation is always looking for those protection systems and verification that they are in working order. Pressure relief valves, fuses, thermal-cut-off, warning systems, shut down circuits are all included in the RCM review.

Safety Devices Protect the Operator, Machine, or the Entire Factory

- Hidden like a thermal overload switch or an overspeed cutoff.
- You don't know there is a problem until it is too late.
- Accidents do happen.
- Must be part of your RCM analysis.

Cost

In operational failure modes (such as the motor failing), the cost of any PM tasks over the long haul must be lower than the value of the repair and the downtime. If the PM tasks cost $1000 a year and a breakdown cost $2500, and downtime cost $4000 to repair, then the failure must be avoided more than every six years.

Random Enemies

The crazy thing about this study was that among other exciting things when they looked at failures, they found wear and tear or age did not cause most of the failures. Most breakdowns were random. The best example of a random failure is you are driving down the road, and rock breaks your windshield.

No amount of PM would anticipate that or mitigate that kind of failure. Other random events like road hazard tire failures, gunk in the diesel, brake failures (some), some transmission failures (a few) can be somewhat managed but not eliminated by adequate PM service.

Their study found that over 75% of the failures were random. They grappled with how you manage these random failures. Think about it at 36,000 feet with 200 souls on board, and you can imagine the sleepless nights. The coffee maker was less important than the landing gear (to most of us).

QUESTIONS OF RCM

1. What is the item supposed to do and its associated performance standards?
2. In what ways can it fail to provide the required functions?
3. What are the events that cause each failure?
4. What happens when each failure occurs?
5. In what way does each failure matter?
6. What task is proactive to prevent, or to diminish to a satisfactory degree, the consequences of the failure?
7. If no suitable preventive task can be found, what's next?

RCM IS A FIVE-STEP PROCESS

The five steps of RCM cover the seven questions. The process is usually team driven with members from operations, engineering, and maintenance (where there is a significant hazard, then safety or environmental specialists would be included). It is usually facilitated by an RCM specialist with good knowledge of the process and products.

1. Identify all the functions of the asset. At first, this might seem trivial. On a second examination, many secondary features are essential. Functions are divided by primary, secondary, and protective (also called hidden). Each function is defined by a specification or performance standards.

For example, the primary function of a conveyor is to move the stone from the primary crusher to the secondary crusher. The specification calls for 750 tons per hour capacity. Secondary functions include containment of the crushed stone (you don't want pieces falling through the conveyor and hurting someone).

2. The second step is to look at all ways the asset can lose functionality. These are called functional failures. One function can have several operational failure modes. A complete technical failure would be that the unit cannot move any stone to the secondary crusher. A second failure would be it can run some amount less than the specification of 750 tons per hour. A third functional failure is if the conveyor starts moving more than 750 tons per hour and starts to overfill the secondary crusher.

Each secondary function also has losses of functionality. In our example, the conveyor could allow stones to fall to the ground creating a safety hazard.

3. Review each loss of function and determine all the failure modes that could cause the loss. Failure modes are not restricted to breakdowns or maintenance issues. Operational problems, problems with materials, utilities are also considered. In our example, the list might be 20 or more failure modes to describe the first functional failure alone. Failure modes or our rock conveyor include several types of motor failure, belt failure, pulley failure, inadvertently turning the unit off, power failure, etc. Each functional failure is looked at, and the failure modes are defined.

Use some judgment to include all failure modes regarded as probable by the team. All failures that have happened would be covered as well as other likely occurrences. Take particular care to include failure modes where there is a loss of life or limb or environmental damage.

It is essential that the team identifies the root cause of the failure and not the resultant cause. A motor might fail from a progressive loosening of the base bolts, which strains the bearing, causing failure. This would be a motor failure due to base bolts loosening up.

It is crucial to include failure modes beyond normal wear and tear. Operator abuse, sabotage, inadequate lubrication, improper maintenance procedure (re-assembly after service), would be considered.

4. What are the consequences of each failure mode? The effects fall into four categories. These include safety, environmental damage, operational, and non-operational. A single failure mode might have consequences in several areas at the same time. John Moubray, in his significant book Reliability-centered Maintenance II, says, "Failure prevention has more to do with avoiding the consequences of failure then it has to do with preventing the failures themselves."

The consequences of each failure determine the intensity with which we pursue the next step. If the effects include loss of life, the failure mode must be eliminated or reduced to improbability.

A belt failure would have many consequences, which would include safety and operational. A failed belt could dump stone through the conveyor superstructure hurting everyone underneath. The failed belt would also shut down the secondary crusher unless there is a back-up feed route.

The failure of the drive motor on the conveyor will cause operational consequences. Operational consequences have costs to repair the fault itself as well as the cost of downtime and eventual shut down of the downstream crushers.

Other failures might only have nonoperational consequences. Nonoperational consequences include only the costs to repair the breakdown.

5. The final step is to find a task that is technically possible and makes sense to detect the condition before failure or otherwise avoid the consequences. Where no task can be found, and there are safety or environmental consequences, then a redesign is demanded.

For example, if the belts start to fail after they are worn to 50% of their thickness, a periodic thickness inspection might be indicated. If the belts fail rapidly after cuts or other damage, then a sensor might catch these problems. In all cases where safety or environmental damage is the primary concern, the task must lower the probability of failure to a very low level.

You Start with a Failure Mode

Let's say the loss of engine power is the failure mode. What are all the things that could cause this loss of power? Some are typical maintenance issues like:

- Bearing failure
- Crank breaking
- Clog in the fuel line
- Camshaft breaking
- And about 100 more

But wait, many of the causes of loss of power are not maintenance at all, like:

- Bird strike
- Run out of fuel
- Pilot/driver reducing the throttle
- Bad fuel

Defense in depth considers the causes and seeks to eliminate all of them. Defense in depth makes the system (truck, airplane, submarine) more robust.

Second, it means applying all the tools in your toolbox toward eliminating or reducing the chance of the bad ones from happening.

Tools include reengineering, refining pre-trip checklists, change to training, enrolling the driver in run time inspection, finding new aftermarket products, new procedures, ensuring procedures are followed, etc. All of this will benefit from specific discussions with all the participants (from mechanics, drivers, washers, fuelers, part room people, etc.).

PMO (PM OPTIMIZATION)

Review of OEM PM Recommendations: Steve Turner OMCS International

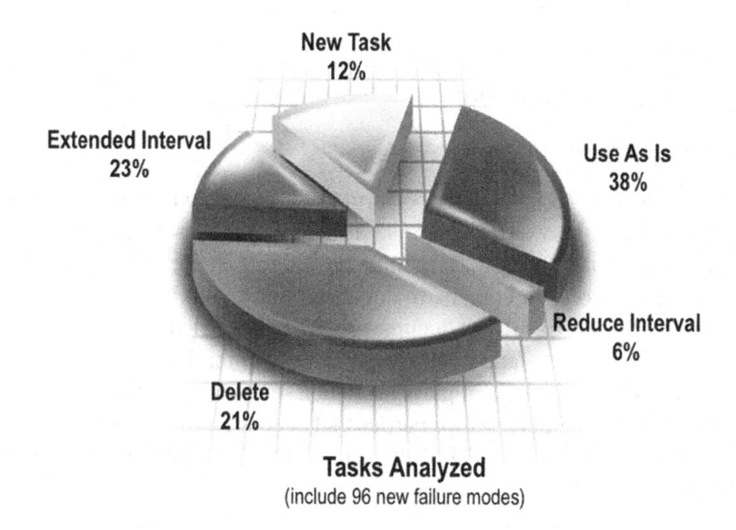

FIGURE 15.2 OEM PM tasks analyzed.

I would like to thank Steve Turner, a professional engineer from Australia and RCM expert, for introducing me to PMO. He developed PMO out of frustration with the application of RCM in mature industries. Much of the material is adapted from his writings.

PMO (PM Optimization) is an offshoot of RCM and recognizes the difficulty (and sometimes futility) of RCM in a mature operational plant. PMO embodies techniques to optimize the PMs to get the most reliability from the least resources.

What would happen if you took the excellent structures of RCM but skipped the function part and looked only at failures that have happened or are likely to happen? Well, if you skipped these steps and added in some common sense, you would end up with PMO. PMO is designed explicitly for mature industries where the opportunity for equipment redesign is limited.

RCM came out of an environment where, if the system was a problem, it could be redesigned. In most factories, buildings, and certainly fleets of vehicles, the equipment is just a given of the equation. Some redesign can be done, depending on the capabilities of the organization, but it is usually very limited in scope. Typically, factories have more capacity for reengineering than fleets or building maintenance departments.

PM Optimization is a best practice that is achieved by:

- Removing or enhancing all vague maintenance tasks (check, inspect) don't add any value or are not cost-effective.
- Replacing calendar-based tasks with run-based, condition-based, or run to failure were feasible and cost-justified.

- Eliminating duplicate PM is where different people or groups are performing the same PMs to the same assets.
- Assigning tasks appropriately between maintenance and operations.
- Making PMO a living program updating as needed.

Nine Steps to PM Optimization

1. Task Compilation

Create a catalog of all tasks already performed by anyone who has contact with the equipment. Usually, this would comprise all current PM tasks (of course) but also tasks done by machine operators, quality personnel, cleaners, calibration departments, safety inspectors, and others. Include every task with frequency.

Some of the PM efforts are not documented and are carried out on an ad hoc basis by tradespeople, operators, and contractors. Thus, a compilation of the written task lists from the CMMS (if one is in place) can be a starting point, but it is not enough. Direct interviews are needed with all parties that encounter the equipment. Generally, the data is collected into a spreadsheet, which facilitates further steps with the ability to sort the data into different columns.

Create a catalog of all tasks already performed by anyone who has contact with the machine.

Task	Trade
Task 1	Operator
Task 4	Maintenance
Task 2	Operator
Task 7	Maintenance

2. Failure Mode Analysis

In RCM, a great deal of thought and time go into looking at the functions and the function failure to determine all possible failure modes. In PMO failure mode analysis (FMA), the team works from the accumulated tasks back to the failure mode. In other words, failure modes without tasks are not considered initially (they are considered later). This one difference cuts the time of the project dramatically over an RCM project of the same size. The task compilation is the basis for this part of the project. The question to be answered is, what failure mode each task is addressing? A cross-functional team is best for this kind of analysis.

Task	Trade	Cause
Task 1	Operator	Failure 1
Task 4	Maintenance	Failure 3
Task 2	Operator	Failure 2
Task 7	Maintenance	Failure 1

In other words, failure modes without tasks* are not considered at this time (see step 3). The question that is answered here is, what failure mode each task is addressing?

3. Rationalization and FMA Review

Rationalization is simple, put like causes together. Remove tasks with no failure mode and add tasks where there is an uncovered failure mode.

When all the tasks from all the sources were loaded into a spreadsheet, then sort the spreadsheet by failure mode.

Task	Trade	Cause
Task 1	Operator	Failure 1
Task 7	Maintenance	Failure 1
Task 2	Operator	Failure 2
Task 4	Maintenance	Failure 3
Failure mode without task*		Failure 4
Task 5	Maintenance	NO failure mode

At this point, the team can readily see if there are failure modes covered by duplicate tasks or covered by clearly inadequate tasks.

Equipment history (from CMMS or people's memory) is consulted to make sure that all failure modes are listed. The team reviews all aspects of the asset to determine whether there are significant failures that are not covered by any task. Hidden failures are frequently in this category. Failure four in the chart is a failure without a task associated with it.

4. Optional Functional Analysis

In some analyses, an RCM type of functional analysis and loss of function are indicated.

In some analyses, an RCM type of functional analysis and evidence of loss of function are indicated. This approach can be justified on highly sophisticated equipment where the consequences of failure are severe. In these few events, a sound understanding of function is essential to determining that all maintenance and operational issues are covered.

5. Consequences Evaluation

One of the breakthroughs of RCM is its focus on consequences rather than failures themselves.

Task	Trade	Cause	Effect
Task 1	Operator	Failure 1	Operational
Task 7	Maintenance	Failure 1	
Task 2	Operator	Failure 2	Operation
Task 4	Maintenance	Failure 3	Hidden

Determine consequences from the failures. These categories are analogous to the ESON factors in RCM.

6. Maintenance Policy Determination

This step is the core of PMO. Based on the consequences, review each task. They're a series of questions the PMO team asks about each task.

- Does the task make microeconomic sense? The first determination concerns microeconomics. Is this individual task (labor and materials times frequency per year) worth the cost, given the value of the failure times the probability of the failure in that year?
- Is there a better way to get to this failure mode? In some circumstances, the introduction of quick condition-based monitoring would save overall time and money. The corollary is, would this task respond to simplification of the technology?
- Eliminate tasks that serve no purpose. Determine which tasks can be set up at lesser or greater frequencies to increase effectiveness.
- There is always an issue of what data to collect and to what end. The analysis at the task level answers this critical question. What data should be obtained to be able to predict the life of this component more accurately?

7. Grouping and Review

This step is efficient in that it looks at the tasks that are left after duplicates, and uneconomical ones were eliminated and divvy them up based on the facts of your operation.

This step looks at the tasks that are left after duplicates, and uneconomical work was eliminated and divvied them up based on the facts of your operation (proximity, frequency, the skill required). Questions such as do operations get all the daily tasks and should the night shift be given accountability for this specific asset are answered.

8. Approvaland Implementation

All parties must be informed about what changed and why. All parties must be informed about what altered and why. All stakeholders are involved in this step. The change must be communicated to both maintenance and operations personnel and staff. The more complex the operation is, the more critical this step becomes (and the more difficult).

9. Living Program

Turning a PM program into a living program requires time and patience. The less wasted PM will mean immediate freeing of resources. As these resources are reinvested to clean up the backlog, and the effective PM strategy takes hold, the number of breakdowns decreases. As the number of failures decreases, more resources are freed up and can be used to accelerate the whole program.

Other steps can now be taken to improve the whole maintenance effort. In this context, the changes currently contemplated will make a difference.

First, go after Low Hanging Fruit! PMO is a fruit magnet!

16 PM for Supervisors *What You Must Know*

Your PM program is the core effort to anticipating failures and extending the life of your assets. When we discuss reliability from the perspective of the maintenance department, almost all the tools and tricks are contained in the PM task lists. At the same time, there are high tech efforts to supplant PM; that event will be years in coming. BAE (Become An Expert) your success depends on it.

SOME PM TERMS

Asset, **unit**, **equipment**, and **machine** are interchangeable. In this case, all mean the basic unit of maintenance

Breakdown or emergency work: Any work that interrupts and displaces scheduled work

Corrective Maintenance CM: Maintenance work identified by an inspection that is plannable and schedulable before failure.

Failure Experience (both frequency and severity) feeds back into a task list

Frequency: How often to perform tasks on the task list?

PM: Preventive Maintenance: Not Panic Maintenance, Percussive Maintenance, pencil whipped maintenance, project management

P^3 = Prevent failure, Predict failure, Postpone failure

PdM: Predictive Maintenance-inspection with an instrument

PPM or **P/PM** is Preventive and Predictive Maintenance

Task list: a list of all the tasks to be done.

TLC (tighten, lube, clean) Basic care.

PM Report Card

How good is your department in PM? Give yourself grades in these aspects of Preventive Maintenance

Preventive Maintenance consists of following activities on a task list:

- PM includes all types of scans, monitoring, and detective activities. These include visual (and feeling for heat and vibration), listening, smelling inspection, taking measurements, inspecting parts for quality, and analysis of the oil, temperature, and vibration. Recording all data from the predictive activity for trend analysis
- Lubricate, servicing automatic lube systems, topping off reservoirs

PM activity	Your Grade
Inspection (human senses)	
Ongoing training and documentation of what to look for	
Hi-tech inspections (predictive maintenance)	
Cleaning program	
Tightening bolts	
Lubrication program	
Visual workplace – placards to tell PM person what to do and show it	
Getting operators involved in PM	
Checking the operation of the unit	
Minor adjustments	
Take and trend readings	
The traceable link between PM findings requiring action and the written corrective maintenance work order	
Interview operator before PM	
Checking unit history before PM	
Periodic Analysis as part of PM	
Excellent record-keeping	
RCM or PMO analysis	
When failures occur: Failure analysis	
Use of planned replacement, overhaul, or planned discard	
Use of wired and wireless sensors	

FIGURE 16.1 Report card for PM activities.

- Adjustment, cleaning, tweaking
- Short or minor repairs up to 30 minutes in length
- It also includes PCRs, replacement of components before they fail, replacement of filters, belts, etc.

The complete PM system consists of additional essential activities:

- Writing up any conditions that require attention (conditions which will lead or potentially lead to failure).
- Write-ups of machine condition, log entries
- Scheduling and doing repairs written up by PM inspectors (this is critical)
- Using the frequency and severity of failures to refine the PM task list
- Maintaining a record-keeping system to track PM, failures and equipment utilization
- Continual training and upgrading of inspector' skills, improvements to PM technology

PM (Preventive Maintenance) Can Come from Other Techniques

A complete PM cycle is a highly efficient tool for organizing all maintenance activities. Procedures and checklist are required. The results of RCM and PMO are more concise tasks, frequencies, and choices for the PM program.

Mandatory PM (as opposed to discretionary PM, which is driven by downtime avoidance and cost avoidance) is required by one of the following:

- Federal agencies
- OSHA, EPA, DOT, NRC, FERC, etc.
- Insurance Companies
- Operating Licenses
- Protect personnel or citizenry
- State or Province
- Pressure vessels and boilers
- Manufacturer warranties
- Verification and documentation

These are PMs required by law, permit, an insurance company, or regulator. Not doing these PMs could result in fines, loss of insurance coverage, or even the loss of the ability to operate the asset. In the worst-case, not following these requirements could be a crime.

The task list should capture information about or direct the attention of the inspector toward critical wear areas and locations. If you are inspecting an expensive component system, many inspections might go by without any reportable changes. Depending on the economics, you may want to continue to check to capture the difference when it happens. Always continue to inspect life safety systems.

You must design standards to increase and decrease the number of tasks based on the failure history. Some organizations use the rule that if they don't get a reportable item every other PM, they must be inspecting too frequently. Task list items that are directly concerned with life safety items are not included in this analysis. Do not include statute-driven inspections (boilers, sprinklers, etc.).

Four Dimensions:

How to view PM (Preventive Maintenance) and PdM (Predictive Maintenance)?

- Engineering
- Economic
- People-Psychological
- Management
- Possibly a 5th dimension which is statutory (already mentioned)

Don't forget Iatrogenic failures: Every time you touch an asset, you run a real (but small risk) in making it worse. So, touch it as little as possible.

Tests make mistakes (readings, scans, etc.) There will be false positives and false negatives. All tests (medical, engineering, chemical, etc.) will sometime return incorrect results. The more sensitive the test, the higher the number of mistakes.

A mistake that says the machine is OK when it is not OK is called a "false negative." An error that says a device is not ok when it is ok is called a false negative.

With that:

- Can you sell the benefits to the money people, bosses, and skeptics?
- Do you have confidence in the relationship between the deterioration, damage, and defects, and the readings?
- Can you be patient?
- Do you have confidence in the results?
- Do you have confidence in the technology? Do you have confidence in the process? Can you take the heat when mistakes happen?

PM, PDM WORKFLOW (THE BIG PICTURE)

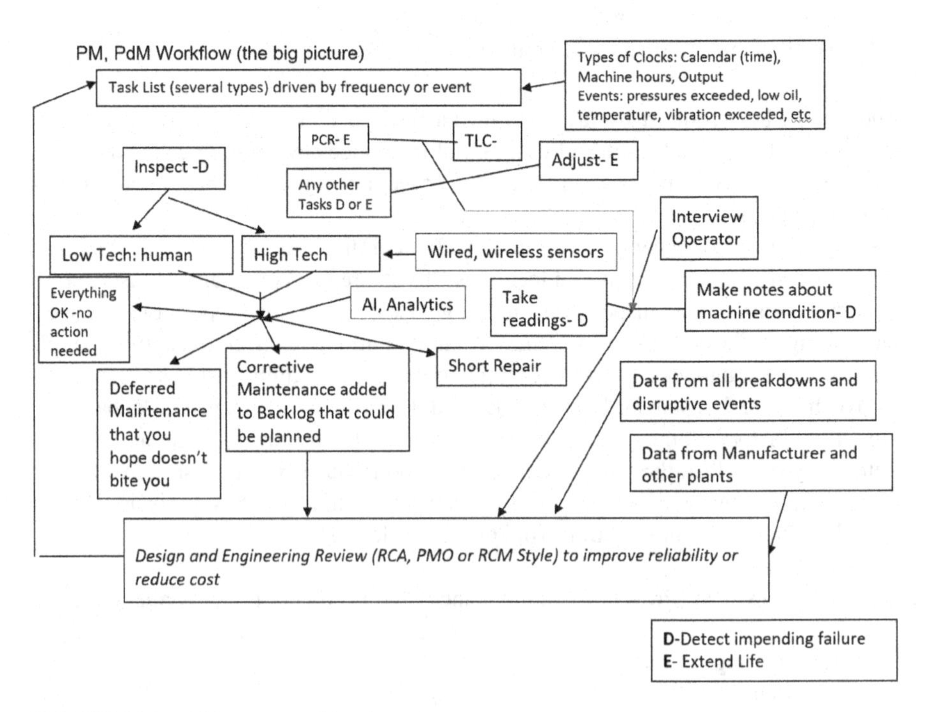

FIGURE 16.2 PM program (big picture).

Practice maintenance reduction

Any activity designed to eliminate maintenance, reduce maintenance, automate maintenance, simplify maintenance

- ME - Maintenance: Elimination get rid of unnecessary tasks
- MP - Maintenance Prevention: Control the sources of dirt, keep the grease inside

- MI - Maintenance Improvement: Better technique, reduce time, and better maintenance products
- All forms of defect elimination!

HOW WE LIKE TO SEE IT WORK!

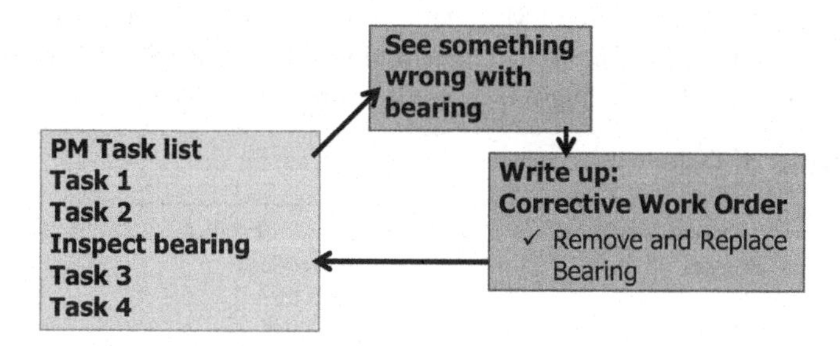

FIGURE 16.3 How PM works.

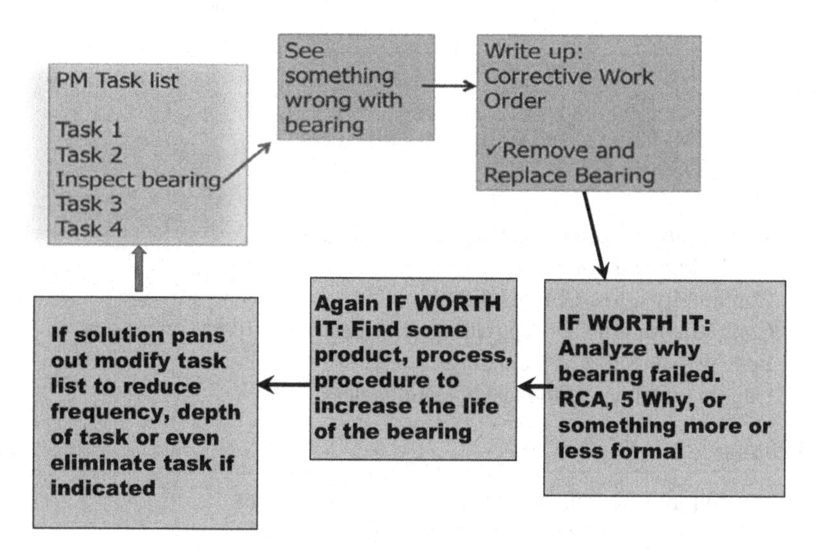

FIGURE 16.4 How we would like to see PM work.

PM Basics

The task list in its highest form, it represents the accumulated knowledge of the manufacturer, skilled mechanics, and engineers, in the avoidance of failure.

Task List Remember P^3

- Prevent Failure
- Postpone Failure
- Predict Failure

The task list is the **heart** of the PM system. It reminds the PM inspector what to do, what to use, what to look for, how to do it, and when to do it. Without having a well-designed task list and following it, you can't improve the PM effort.

Sample Types of Tasks

Prevent, postpone:	Predict
Clean	Inspect Scan
Empty	Smell for
Tighten	Take readings
Secure	Measure
Component replacement	Take a sample for analysis
Lubricate	Look at parts
Refill	Operate
Top-off	Jog
Perform short repair	Review history
	Write-up deficiency
	Interview operator

Also

- Track PM, failures, and equipment utilization
- Predictive activities.
- Short repairs
- Writing ups—Corrective Maintenance (CM)
- Scheduling and doing CMs
- Continually refine PM task list
- Continual training and upgrading of inspector's skills
- Improvements to PM technology
- Ongoing analysis of the task's effectiveness

PM cannot:

- Put iron into an underpowered or too-small machine
- Deal with mis-operation and driver abuse
- Deal with incompetent mechanics
- Deal with random failures
- Deal with bad fuel, wrong lubricants
- Deal with cheap "will-fit" or counterfeit parts

Types of Task Lists

Unit-based: This is the standard type of task list where you go down a list and complete it on one asset or unit before going on to the next asset. The mechanic would also correct the minor items with the tools and materials they carry.

Variations:

- Short repair: done at the time as the PM
- Gang PM: A whole team does the PM at one time
- TPM approach: Operators do the PM

String-based: Your string-based list will perform one or, at the most, a few short PM tasks on many units at a time. Examples lube routes, vibration routes

Advantages: Easy to teach, an excellent way to show an area, can be efficient

Disadvantages: Redundant, no one is accountable, boring

Condition-based PM (CBM): This is a PM mode made immensely more popular by computerized control systems in vehicles, buildings, and factories. Modern SCADA systems, building management, and vehicle computers can feed data to a condition-based maintenance decision engine.

Conditions might include:

A machine cannot hold a tolerance
A boiler pressure gets too high
A low oil light goes on
The pressure drop across a filter exceeds a limit
Amp readings have been trending up

Once the condition goes out of bounds, initiate the corrective action.

Permanent Sensors (IIoT): Biggest Trend Now!

Low cost
Long battery life (some even have built-in microgenerators)
Several readings together
Continuous readings (1 per second, minute, hour, day, etc.)
Wired and wireless upload (no visit required)

IIoT Sensors

These sensors have been developing quietly for decades. Once consumer electronics (smartphones, wearables, game controllers) became a significant business, sensor development took off to reduce size, power, and cost.

- RFID
- Motion
- Cameras
- Hi-speed capture
- Temperature
  - Duct temp
  - Water temp
  - RTD high, low

- Vibration (accelerometers)
- Sound level
- Mechanical stress
- Vibration
- Dry contact
- Open/closed
- Water present
- Voltage
- 0–5 V
- Voltmeter
- Volt detect
- Conductivity
- Pressure
- Spectrometer
- Resistance
- GPS
- Amps
- Amp meter
- Current detect
- Tilt
- Impact
- Movement
- G force (snapshot)
- Weather
- Colorimeter
- Turbidity
- CO_2
- PH
- Light

Staffing the PM Effort

"A successful PM program is staffed with sufficient numbers of people whose analytical abilities far exceed those of the typical maintenance mechanic." (from August Kallmeyer, *Maintenance Management*)

- PM has two sides: TLC and Inspection
- What kind of person do we want for this job?

Six Attributes of a Great PM Inspector

1. Inspector can work alone without close supervision. The inspector has to be reliable since it is hard to verify the work.
2. The PM inspector is interested in the new advanced predictive maintenance technologies. Trained-in techniques of analysis and the use of these modern inspection tools is essential.

3. The PM inspector should know how to (and want to) review the unit history and the class history to see specific problems for that unit and for that class. They should also be the type of person who will fill out and complete the paperwork.
4. A mechanic is re-active in style. A PM inspector is proactive in manner. In other words, the inspector must be able to act on a prediction rather than react to a situation. He/she is primarily a diagnostician, not necessarily a "fixer."
5. Because of the nature of the critical wear point, the more competent the inspector, the earlier the deficiency will be detected. The early detection of the problem will allow more time to plan, order materials, and will help prevent core damage.
6. PM inspectors should be full time and be segregated, if practical, from the rest of the maintenance crew. In some large operations, the PM group receives a paid premium and may have a different style uniform. In any facility, the PM inspectors should represent 10–20% of the whole crew.

Supervisors Take Note!

You are accountable for the PM tasks being done as designed. How to ensure the PMs are done as designed?

- Inspect the work on a random basis
- Does the inspector know how the PM activity fits into the overall scheme?
- Drag your top management down to the bowels of the facility and have them address the maintenance crews about the criticality of PM and its impact on output, safety, etc.
- Present the job as necessary.
- Have a display of PM accomplishments in a public area.
- When uptime is excellent, make it a practice to send out letters of commendation to the PM crew
- One of the most important things you can do to ensure the cooperation and complete work is to let your PM mechanics themselves design elements of the system and tasks themselves.
- One typical hole is a lack of specific skills. Be explicitly sure the PM people are fully trained.
- Follow the PM person from time to time. Record their ideas for improvement and make improvements.
- If training and testing is involved, prepare and give out certificates of competence
- Improve the relationship between the mechanic and the maintenance user.
- Make it easy to do tasks by reengineering them.
- Simplify paperwork.
- Improve accountability by mounting a sign-in sheet inside the door to the equipment.
- Make PM a game.
- PM professionals like new, better toys (sorry better tools, not toys).

- In any repetitive job, deal with the boredom. Consider job rotation, reassignment, project work, office work like planning, design, and analysis to improve morale.
- Be sure that PM is part of the regular reporting up the ladder in the company. Train managers to ask questions when the numbers change.

Short Repairs and High Productivity

Short repairs: Repairs that a PM or route person can do in less than 30 minutes with the tools and materials that they carry. These are complete repairs and are distinct from temporary maintenance.

Do short repairs at the time that the PM inspector is in front of the asset. These are repairs where the spare parts are right there, and the inspector does not have to go to the stockroom. Finally, the mechanic is already carrying the tools needed to do these repairs.

There are three rules for short repairs

- You must set a maximum time depending on the size of your facility and your type of equipment. Usually, the limit is 15 minutes, but restrictions as long as an hour are common. Exceptional circumstances (when the item is far away) can allow more extensive short repairs.
- The repair must be able to be done safely with the tools that the PM person has with them.
- The PM mechanic must be already carrying any necessary materials or parts

The short repair is charged to corrective maintenance if this is possible.

Design a short repair cart (this also can be used for route maintenance)

One way to extend the scope of short repairs is to supply the PM people with carts. Since two of the rules involve tools and spare parts, the solution is to design a cart or mobile job box with the commonly needed tools and spares. The cart of tools available for short repairs will carry supplies appropriate to the plant and equipment they must maintain. For building maintenance situations, for example, the PM maintenance person might be equipped with:

- Hand tools including screwdriver set, pliers set, claw hammer, cutters, Allen wrenches, vice grips, keyhole saw, hack saw, tape measure, utility knife, pipe wrenches, set of files, rasps, a good flashlight, batteries, etc., plus a stepladder tall enough to reach the ceiling
- Electric tools such as electric drill and bits, drop light, skil saw, and angle grinder
- Cleaning tools: straw broom, whisk broom, dustpan, trash bags, mop, wringer, bucket, rags, shovel, sponges, 5-gallon bucket, spray bottles, razor blade scraper, and steel wool
- Cleaning supplies: furniture polish, all-purpose cleaner with TSP, spray deodorizer, spray tile cleaner, wax, wax applicator, wax stripper, toilet bowl cleaner, oven cleaner, metal polish, nonabrasive cleanser, rags, and paper towels
- Silicone spray lube, WD40, spray paints, spray zinc, standard off-white latex paints (or standard colors), brushes and rollers, joint compound, spackle knife, spackle tape, contact cement, latex, and silicone caulk and gun

- Variety packs of fasteners, variety of nails, small hardware items, duct tape
- Elect: replacement LED tubes and lamps, switches, outlets, switch, outlet and blank covers, electrical tape, fuses, fittings, outlet tester, neon tester, door hardware, locksets, doorbells, transformers, wire, smoke detectors, batteries, tags for writing dates of installation and testing
- Window hardware, floor and ceiling tiles, threshold and entrance strips
- Bug bombs, insecticide spray, can hornet/wasp killer, and roach/ant traps
- Faucet washers and seats (seat tool), kitchen and bathroom faucets with flex lines
- Toilet parts, closet seals, toilet seat parts, and closet snake

Adding to the PM Cart

Each cart or each area has a cart Inventory list. The cart should always carry these items. Important: the last daily task is for the cart to be replenished, batteries charged, and tools cleaned. Study the maintenance log and the corrective work orders. Add items based on jobs requested.

Meet periodically with the PM crews to discuss jobs completed and jobs that could not be completed. Adjust the cart based on these discussions.

Allow the individual PM personnel to add things to the cart on their own. Again, at the periodic meeting, discuss the personal additions to see if they warrant adding to the cart inventory list.

The key to these carts is discipline. All tools and unused materials must be put away into the same places, pockets, drawers, and cabinets each time. Care is taken to clean, lubricate, charge batteries, and generally care for the tools every PM Day.

Route Maintenance

This is a related strategy commonly followed by facilities maintenance departments. In route maintenance, you set up a route that you visit every periodically (like weekly, biweekly, monthly, etc.). The service technician visits each area regularly and does all the minor service and PM for that area.

Access to Equipment for a PM

Problems with Production: Not Getting the Equipment

PM is all well and good. What if production won't give us the equipment?

Common complaint: I have a PM schedule, but production will not give me the machine. A week or a month later, the device fails, and I get blamed!

There are two kinds of access problems:

Political access problems are problems that stem from political reality. The equipment is not in use for 24 hours, seven days. But it may seem to be in use whenever you want it for PM.

- Go back to the planning department to discuss requirements. The goal is to set-up a 12-month schedule for PM with the MRPII schedulers.
- Your PM requirements are distrusted. You may have to haul in respected outsiders.

FIGURE 16.5 Display case for broken parts.

- Always collect broken parts and build a display.
- Use any production and downtime reports now in circulation; highlight downtime incidents that could have been avoided by PM effort.
- Most importantly, conduct your business with Production with integrity.

When It Positively Cannot Be Shut Down?

We call these engineering access problems. Engineering access problems are easy to spot. These access issues stem from equipment that cannot be taken out of service because it is always in use.

A partial antidote for both political and engineering access problems might be through non-interruptive maintenance. Non-interruptive maintenance is maintenance that can be done while the machine is running. Thus there is no interruption.

Interruptive/Noninterruptive

Interruptive tasks	Noninterruptive tasks
Check coupling tightness	Scan all vibration points
Verify tightness of electrical connections	Check anchor bolt tightness (sometimes)
Jog machine to see alignment	An infrared scan of connections
Clean out jaws area	Sweep up around machine (sometimes)

In some cases, the tasks neatly divide into two categories. If you can't have control of the asset, the non-interruptive list might a good approach to PM. You end up with two task lists for each asset in this category.

Advantage: Reduced machine downtime

Disadvantage: Slightly less productive since the machine requires at least two trips

The Next Logical Step: Reengineer the Tasks

In some cases, we can reengineer the task so that it can be done safely without machine interruption. The easiest example is re-piping lubrication lines to a safe area. Other ideas include putting windows to see inside equipment safely or adding sensors to read out the conditions inside without opening it up.

Another approach is the substitution of a noninterruptive task for an interruptive one. This substitution may require the adoption of technological solutions. In the task list above, consider the two tasks:

Interruptive task	***Noninterruptive task***
Check coupling tightness	*Scan all vibration points*

17 TLC (Tighten, Lubricate, Clean)

PM is two-sided. One side is the detection of defects, damage, and deterioration and correcting it before failure. The other side is postponing or preventing the damage, deterioration, or defects in the first place. TLC (Tighten, Lubricate, and Clean) will push off the failure to the future. TLC is one of the few activities that, by itself, will postpone failures.

TLC means tender loving care. When we apply this to machinery, we get **T**ighten, **L**ubricate, **C**lean. Keeping equipment trim and clean will extend the life and reduce the level of unscheduled interruptions. This approach or strategy is appropriate for all maintenance departments, even those with no support from top management or maintenance customers. **TLC is the simplest way to reduce breakdowns.**

Lesson 1 in the new world of hi-tech maintenance

- No matter how smart your computers are, the physics-engineering of the machine, building or other physical asset is the same.
- Most of your actionable breakdowns are the result of defects, i.e., damage or deterioration.
- Three sources are causes of most failure modes
  - Loose bolts
  - Dirt and contamination
  - Lubrication issues

TLC can impact other costs. One firm reduced electric usage by 5% through effective lubrication control. TLC also makes assets last longer by reducing the cost of replacements.

TLC (TIGHTEN, LUBRICATE, CLEAN)

Don't make this mistake: Basic maintenance tasks are sometimes missing in favor of "hi-tech" and more fun activities.

Bolting (Tighten)

"Bolts are tightened by applying torque to the head or nut, which causes the bolt to stretch." (Machinery's Handbook)

Misconceptions

Using a torque wrench in infallible. Not always because of friction. Remember, the goal is to stretch the bolt. The bolt's stretching clamps the joint. If there is rust, dirt, the torque will be greater to achieve the same level of elongation, and if there is grease, the torque required will be greater to make the same elongation.

It doesn't matter what the joint looks like when you pick a torque setting. Different joints require different amounts of torque. A joint in tension requires a different torque setting than a joint in shear. A joint in compression has significantly lower torque requirements then either of the others.

All bolts of size should be torqued the same. Bolts come in grades. The range of strength between a grade 1 and a grade 8 is almost 4 to 1. That may mean that the torque to stretch the bolt could vary as much (depending on the application).

Once you properly torque the bolt, you're done. It is a well-known problem in mobile equipment that bolts loosen up in the first 500 miles or 25 hours and should be re-torqued.

No problem with a missing bolt if there are others intact. Loose or missing bolts are a significant source of breakdowns. Even a single missing or loose bolt might cause a failure. While properly engineered joints are designed with structural redundancy, each fastener is essential.

FIGURE 17.1 Match lines to visually inspect for tightness.

An idea for action: The most straightforward technique is to scribe a line on the nut and the machine frame when the nut is tightened correctly. This scribed line will stay intact (a single line) if the nut doesn't move.

Good bolting practice takes a while to teach and is not necessarily intuitive

Lubrication

Failures to lubricate are always the result of several factors. A leading cause is where the grease fittings are too hard to get to or there are just too many points. Other factors include too many different lubricants used, not enough time allowed, lack of standards, and a lack of motivation of the worker.

Don't assume that, even journeymen, mechanics are experts in lubrication.

Tests and Certifications in Lubrication

In the US, there are tests, training, and certifications for lubrication expertise.

Mistakes

Mistakes in lubrication can be devastating. A mistaken lubricant could be spread to all machines in an area in one afternoon through the lubrication route.

Too many lubricant choices lead to problems. One smart strategy is to consolidate lubrication so that you use as few as possible (even standardizing on "better" lubricants).

Clogged, dirty, or broken lubrication fittings compromise the whole effort. Initial cleaning should highlight these issues and correct them.

Twenty-firsty-century lubricate. The operator is now able to listen to the "acoustic" bearing noise through the headset,

- Hear problems associated with a lack of or too much lubricant.
- SKF and FAG suggest that less than 5% of bearings reach their engineered L10 lifecycle.
- They report that as much as 80% of bearing failures are directly attributable to poor lubrication practices
- Use Case: One Day Free Acoustic Lubrication Workshop—by SDT, Ludeca

Have you considered Automated Lubrication Equipment?

SKF use case example

- **$147 US**
- Filled with high-quality SKF greases
- Temperature independent dispense rate
- The maximum discharge pressure of 30 bar over the whole dispensing period
- Dispense rate available in various settings
- Available with non-refillable cartridges in two sizes: 120 ml and 380 ml
- Easy to install and use
- High water and dust protection (IP = 67)
- Extensive range of accessories

Try this 1-hour Lubrication Check List (partially adapted from TPM Development Program)

1. Are lubricant containers always capped?
2. Are the same containers used for the same lubricants every time, are they adequately labeled?
3. Is the lubrication storage area clean?
4. Are adequate stocks maintained?
5. Is the stock area appropriate in size, lighting, handling equipment for the amount stored?
6. Is there an excellent long-term relationship with the lubricant vendor?
7. Does the vendor's sales force make useful suggestions
8. Is there an adequate specification for the frequency and amount of lubricant?
9. Are there pictures on all equipment to show how, with what, and where to lubricate and clean?
10. Are all zerk fittings, cups, and reservoirs filled, clean, and in good working order?
11. Are all automated lubrication systems in good working order right now?
12. Are all automated lubrication systems on PM task lists for cleaning, refilling, inspection?
13. Do you have evidence that the lubrication frequency and quantity are correct?
14. Is oil analysis used where appropriate?

FIGURE 17.2 Quick audit of lubrication area.

Save money by rethinking

- The strategy is to use oil analysis to see if the oil is still in good shape, especially if the reservoir is large. That way, extended drain intervals translate to less oil used and less waste oil generated.
- The second part of the equation is to either mount a bypass low micron filter on the equipment or purchase a filter cart and periodically (thoroughly) cleans the oil in place.
- Consider the "point of use" color-coding (and geometric coding—for color blind people).
- Keep your lubricant storage areas clean, well lit, and well ventilated.
- Do not store barrels of oil outside.
- Consider automated lubrication equipment.

CLEAN

Cleaning builds a relationship between the equipment and the cleaner. The connection is called ownership

No One Likes Cleaning!

In many places, when operations issue a Permit to work or Lock-out Tag-out, they are required to clean, cool down, and decontaminate the machine before releasing it to maintenance.

When this does not happen, the pressure is still on maintenance to make the repairs. As the supervisor, your role is first to advocate that operations do their job. Second, you must engage your team to do the cleaning themselves. To capture the effort, add a cleaning task to the work orders, and be sure everyone enters the cleaning time.

Cleaning: Cleaning Program Checklist

1. Cleaning the main body of machine checking and tightening bolts
2. Cleaning ancillary equipment checking and tightening bolts
3. Cleaning lubrication areas before performing lubrication
4. Cleaning around equipment
5. Treating the causes of dirt, dust, leaks, and contamination
6. Improving access to hard to reach areas.
7. Developing cleaning standards

Even with a Million Dollars of Tech

There still is heat, dirt, vibration, loose bolts, damage, deterioration, housekeeping issues, and other weirdness.

Tiny particles of dirt (0.0015) compared to Internal clearance (0.00016). Even human hair (0.004) is much bigger than dirt!

Dirt: Public enemy #1
Heat: Public enemy #2.

FIGURE 17.3 Public enemy #1-dirt, Public enemy #2-heat.

Keep Area Clean

Keeping it clean is not only a PM issue. Cleanliness also promotes safety and positive morale. Cleanliness is vital in rebuilds, significant repairs, and even small repairs. Any mechanic in the business for a length of time can remember a perfect repair gone wrong because of dirt.

Initial cleaning and machine review: (it would be a win-win if maintenance and production did this together)

- Goal: Restore to original condition: This will halt accelerated deterioration
- Create defect list (and hang tags to make it easy to locate)
- Identify difficult work areas
- Identify lubrication points
- Create a lubrication plan
- Review existing preventative maintenance
- Ask any questions
- Treat the causes of dirt!

Clean to Inspect

- If you clean your car with an automatic car wash, you do not examine the vehicle at the same time.
- In hand cleaning, you inspect as you clean. You then can see defects like loose bolts, scratches, and dents. These can then be fixed.
- The same idea works with machinery. Clean it to inspect it.

18 Work Execution Management

The core of the supervisor's job is managing work execution.

SOURCES OF WORK

Where Does Your Work Come From?

Several sources are internal (PM tasks, corrective maintenance), and the rest are external to the maintenance department. A study of your information flow might show where work requests could fall through the cracks. It would also show where you are not capturing the minimum amount of information to dispatch maintenance workers efficiently.

Process: All service requests should converge in the maintenance control center and be reviewed by a single person or group. If you use a team structure, then the team point person on planning and scheduling should review the requests. Emergencies should be handled directly (through the supervisor). The goal of the system is to serve the customer efficiently with minimum overhead while maintaining control.

The Function of Job Control

All services requested from all sources go through the inbox. The inbox includes incoming e-mail, phone messages, notes, radio calls, from all maintenance users and PMs from CMMS. Time/date stamping should take place on all jobs entered into the inbox. The inbox should be scanned frequently for high priority work that no one mentioned.

Triage: Emergency jobs are dispatched at once to you or a designated tradesperson on the floor. Record the time of dispatch. After the true emergencies are processed, the rest of the jobs sent to planning for planning.

Process: The next step is vetting. Should this job be done at all? Is the job justified? Is it a duplicate? Is the job a small part of a large job that is already on the books?

UNPLANNED WORK – BREAKDOWN

Firefighting, emergency, breakdown is one of the major concerns of the supervisor.

You are loved or hated by operations based on how you respond to these urgent requests. Your attitude, professionalism, and possibly your future depend on your perceived approach to emergency work.

Yet being good at emergency work does not mean you are a great supervisor. It is like a parent of a hungry child. You might get points from the child for always responding with ice cream and other treats, but you are failing at the parent job if you don't offer a balanced diet.

Your primary job as a supervisor is to ensure the company gets maximum value from the assets it uses. Sometimes that means delaying an unimportant emergency to deal with an essential job to avoid failure in a critical asset.

Ideas about Firefighting Effectiveness

On the floor outside of each of the doors of the fire trucks were a set of boots and a pair of fire fighter's overalls set-up so that you could step into the boots and pull up the overalls. The jackets, helmets, air packs, and entry tools were on hooks in the truck. Could you imagine the cost in lives and property damage if each firefighter had to stand in line at the parts window for his/her helmet or air pack?

Have you ever been to the emergency room in a hospital? When someone comes in with a broken leg, the staff rolls over the orthopedic cart with all the tools and materials for broken bones. They cannot fix everything with what is on the cart, but they can stabilize almost everything and fix upwards of 90% of the emergencies.

Every area of our lives where quick response is essential follows some basic best practices:

1. Have the tools needed for 90–95% of the possible incidents in a cart, truck, box, or another isolated storage area. The toolbox should be mobile and easy to get to the breakdown. In a building, a cart is okay, but in a large industrial site consider footlockers, toolboxes that can be thrown onto the back of a truck or a complete truck is more desirable.
2. The tools are put away into the same places, pockets, drawers, cabinets each time. In a breakdown, everyone must know where everything is. That is why the drawer of all of a hospital's code carts (used when someone stops breathing or their heart stops) always has the same tools or drugs. This practice is so vital in a hospital that there are periodic task forces that review and redesign the standard layouts as better technologies and devices become more commonplace.
3. Tools are cleaned, lubricated, batteries are charged. Tools are cared for after the crisis is over. There is nothing more frustrating than being in the middle of a repair and having a dead battery on a needed screw gun, meter, etc.
4. Predict the types of parts and materials needed and build-up a cart, box, vehicle with the materials required for 90–95% of the quick repairs. Alternatively, consider locked cabinets in each area with critical spares. The parts and materials issue in a typical maintenance setting is far bulkier and more complex than in a hospital (humans come in many styles but only two models, unlike machines!).
5. Replenish the parts used after the crisis is over. In a hospital, a nurse is assigned to inventory and replenish the cart after each use.
6. Create a work order of what was done, what was used, for the records. Part of the job is to clean and put away the tools, replenish the components, and fill out the paperwork.

7. Consider having a meeting a day or two later to discuss what happened (informal postmortem), what went well, and what needs work for next time. This meeting is not to point fingers but rather to identify what worked and what didn't. In a hospital, the M&M committee (Mortality and Morbidity) reviews every death and sickness (that started once in the hospital). The committee looks for lessons learned and if any procedures need to be changed.
8. Spend time training people in response to breakdowns. Let the people that are great at firefighting teach what they do, and how they approach these type repairs. Use the work orders from the last crisis to jog people's memory.

Cart Design

Both the maintenance personnel and management should intensively study the fire fighter's cart. Consider the Phone Company or Gas Company. Great thought goes into how to outfit a service person's truck. Next time you have an opportunity, ask the telephone installer or gas repair person how their vehicle is set-up and why. Apply the lessons to the firefighting cart, van (or even 5-gallon bucket!).

The more often you have the needed part in the cart, the more downtime you avoid and the money you save.

Organizations that are serious about quick response to breakdown do the following:

1. Have meetings on this topic and discuss what happened in the past with their old-timers.
2. Include the maintenance customer in these meetings.
3. Decide who will do what when an asset breaks down.
4. Decide where you will keep the cart and spares.

PLANNED WORK

Job control is the GATEKEEPER to maintenance services.

Backlog

Backlog: The third stop is the backlog. The backlog is the holding tank that helps regularize the flow of work to the maintenance shop. Jobs may enter the backlog in clumps, but they are parceled out in a smooth flow that is directly related to the hours available.

Types of backlog that make up the total backlog

- Planner backlog
- Ready backlog
- Backlog waiting for materials
- Backlog waiting for authorization or approval

- Backlog waiting for engineering
- Backlog waiting for weekend or shutdown

Outside the backlog

- Wish List: The exception is jobs that go directly from the inbox to the wish list (where there is no near-term intention of doing the indicated work). In either case, the workflows out of the holding area of the total backlog. Jobs requiring shutdowns are filed in the 'waiting for shutdown' file.

Some computer systems (CMMS) maintain status codes for each job.

The two most important types for this discussion are the Total backlog and the Ready backlog. When planning is complete, materials are in stock, authorization is obtained, and any barriers to production are handled, the job status changes to **ready**, and the work order joins the ready backlog.

After the job is planned and resources are secured, it moves to Ready backlog. All ready backlog jobs have been authorized, parts are available, priority has been set, and planning (if required) has been done.

Managing the backlog is a crucial way to manage an entire maintenance department.

- Job-status and what prep work is missing
- Even workflow into shops
- Is Crew size, right?
- Metric: Trend Ready backlog and Total Backlog measured in weeks

Once the job is scheduled and issued to the supervisor, it then moves to the open or pending status. By convention, the job is still formally on the backlog list until it is completed. In large jobs, hours may be deleted from the estimate as they are completed.

If planning works correctly, you have an easier job and one that is less stressful and less frustrated. You can assign the task with some confidence that the job can be done without too much hassle. This is the goal of planning and scheduling.

Prioritization

Prioritization is essential where there is more work than workers!

The great dilemma because the priority is used to express the customer's feeling of urgency when it should be used to represent the urgency to the business.

- Customer feelings versus situation on the ground
- Want versus need
- Recognize critically
- The scarce resource must be allocated
- Sets sequence of jobs
- Protects the employees, environment, and the asset
- Many methods boil down to discipline

It is essential to prioritize causes so that resources can be applied appropriately. For example, to decide the priority in a manufacturing environment, we would review the following issues:

Safety and health

- Would this job help avoid failure that could endanger the health or safety of employees, the public, or the environment?

Statutory or mandatory

- Is the job required by law, insurance companies, or risk managers?

Downtime

- Would the failure of the equipment to be worked on-stop production, distribution of products, or full use of the facility?
- Is the equipment to be worked on critical to the operation?
- If the job is not done at all, is plant capacity compromised?

Return on Investment

- Will the job increase output so that the profit will pay for the job in a reasonably short time?
- Is the job necessary to maintain or improve the quality of the product?
- Is the cost of the shutdown more significant than the costs of breakdown and downtime?
- Is the capital investment high in the equipment?

Asset or Engineering

- Will the problem get worse if it is not done?
- If delayed, will the job significantly increase in price?
- Does the average life expectancy of the equipment without this job exceed the operating needs?

Operations

- Can the load be easily shifted to other units, plants, or workgroups?

Maintenance work order is the communication document that circulates essential information around the organization.

There are many methods used to collect information about repairs. Adaptation of outside technology such as bar code scanning, RFID (Radio Frequency ID) chips has simplified, quickened, and made data collection more accurate. Collecting the work order information data has been in the past; the most valuable aspect of having a maintenance management system.

The tool used is the CMMS. Good understanding, knowledge, and skills with the CMMS will make the job easier and make you more productive.

The most significant contribution you can make to the CMMS as a supervisor is to ensure the integrity of the data in the work orders.

19 Planning, Coordination, and Scheduling

Planning and scheduling will make the jobs of maintenance less frustrating, more fulfilling, safer, and quicker. Accurate planning and on-time scheduling will simplify your job. It will also motivate your team and encourage higher quality work.

Are you ready to face the facts?

What really goes on in maintenance…

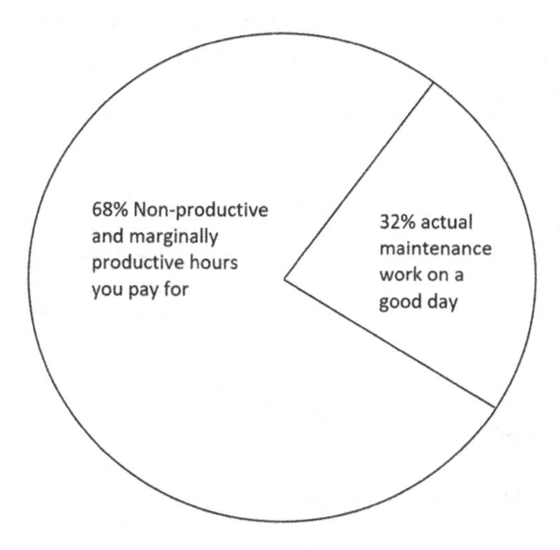

FIGURE 19.1 Productivity of maintenance workers.

This shocking situation is the responsibility of management, if anyone, but usually, it is no one's fault. This most significant barrier to productive maintenance is that the maintenance people are often not prepared to do the job.

People can't effectively work if they don't have EVERYTHING; they need to do the maintenance job. As the supervisor, you must ensure that everyone has what they need, whether your organization has or has no planning and scheduling. Before we discuss productivity, let's define the terms.

In the maintenance world, we collapse the planning and scheduling functions.

Planning: The deconstruction of the maintenance job into the things needed for the successful, efficient, and safe execution of the work. Planning is just making lists. Planning does not move things in the real world.

Coordination: Having the customer involved in picking jobs from the ready backlog for scheduling

Scheduling: The assignment of a time slot to perform the work and the necessary activity to ensure everything on the job plan is available when the job starts.

Execution: Execution is moving and doing in the real world. Everything is on paper to this point.

PRODUCTIVITY OR THE LACK OF IT

Another way to show this is by looking at the details. Hidden in the details of this chart are the barriers to productive maintenance workers. As an example, note that the bulk of the losses occur in getting "stuff" and traveling to get "stuff."

Notice that obtaining tools and materials drops by about 60% when planning and scheduling are introduced. Travel drops by 33%. Why is this so? If the job plan has a BOM (Bill of Material—list of parts) and a unique tool and equipment list, then the maintenance worker can pick up what's needed before leaving for the job.

In an unplanned job, the worker must scope the job themselves and get what they remember. And unfortunately, go back to get what they forgot.

Typical Maintenance Worker's Day: Planning/scheduling Versus nonplanning/scheduling

	Reactive (without planning and scheduling)	**Proactive (with planning and scheduling)**
Receiving Instructions	5%	3%
Obtaining tools and materials	12%	5%
Travel to and from the job (both with and without tools and materials)	15%	10%
Coordination delays	8%	3%
Idle at the job site	5%	2%
Late starts and early quits	5%	1%
Authorized breaks and relief	10%	10%
Excess personal time (extra breaks, phone calls, smoke breaks, slow return from lunch and breaks, etc.)	5%	1%
Subtotal	65%	35%
Direct actual work accomplished (as a percentage of the whole day)	35%	65%

FIGURE 19.2 Maintenance worker's day-time spent.

You might argue that planning and scheduling mean someone must visit the site, so why not the worker? It turns out to be more efficient for a planner to visit and plan. Each dollar invested in planning typically saves three to five dollars during work execution, and the duration of a planned job is commonly only half of that of a job unplanned.

Another way to appreciate the advantage of job planning is to depict what happens within a specific job without a plan. Typically, technicians jump into the work without forethought.

Shortly, they encounter a delay due to the lack of a spare part, tool, or authorization. This sequence may repeat several times before the job is completed. In the planned mode, needs are anticipated and provided for, before a technician is assigned.

PROFESSIONAL PLANNING VS. PLANNING ON THE RUN

Planning on the Run

Professional Planning

By the way, your best workers already do most of the things we will be discussing. Your worst workers probably won't benefit either because they will ignore the planning!

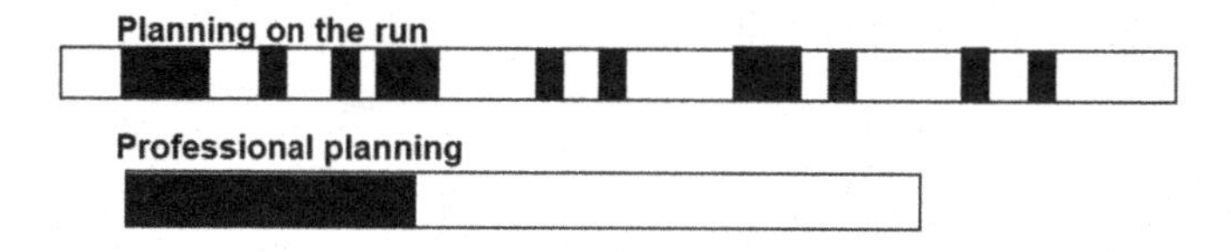

FIGURE 19.3 Planned versus unplanned work.

PITFALL

One of the most significant objections to planning and scheduling is that a planner can never know as much as the worker on the ground. This is an entirely valid point.

It is the supervisor's job to facilitate proper communication with planning.

Why? It will be to your benefit! The best way to improve job plans is for the supervisor, worker, and planner to have excellent communication. Your worker should note any issues with the plan, any ideas they have about improving the plan, and any advice for the next team and get those notes in front of the planner. The planner should correct all mistakes and consider all alternate tactics to perform the job.

TYPES OF WORK

The nature of maintenance enables planning by setting up your crews correctly

PM, PdM, and Routine:	Don't interrupt these people. Let them do their jobs.
Backlog relief:	Working off the jobs on the backlog. Biggest crew.
Emergent:	First responders, fast fixers.

Consider this: Appoint specialists in each type of maintenance as the "go-to" people for each category of work.

If, as a supervisor, you are interested in altering the future workload, then the essential portions of the maintenance workload are reliable PM Service and timely backlog relief. Proper attention here provides asset reliability, minimizes future emergencies, leads to a proactive environment, and offers sustained viability of the organizational entity.

There are far more significant bottom-line contributions to be gained from asset reliability than from "mere" maintenance cost reduction. This approach is also more saleable to the maintenance workforce. By committing themselves to asset reliability, they protect the future of their jobs.

Fear

If the managers focus only on cost reduction, improved productivity from planning and scheduling works the maintenance crew out of overtime and possibly out of their jobs. This is a common fear.

The Elements of Maintenance Planning

Look at the planned work order. While not all these items are needed for every job, every one of them should be considered.

(This is important!) **Planning provides:**
• Safe job steps and procedures
• Hours for the person(s) with the right skill(s) & licenses to perform the job (physically and mentally). And crew size
• Correct parts, materials, supplies, consumables for the job
• Correct special tools
• Adequate equipment for lifting, bending, drilling, welding, etc
• Identification of hazards and mitigation, including Personal Protective Equipment (PPE)
• Specific lock-outs
• Proper permits and Custody and control of the asset
• Safe access to assets (decontaminated, cooled down, etc.), secure work platforms, and humane working conditions
• Up-to-date drawings, wiring diagrams, and other information
• Proper waste handling management
• Testing and commissioning
• Return excess spares, tag and return rebuildables, fill out paperwork, clean up the site!

FIGURE 19.4 Planning provides list of these lists and instructions for these elements.

If you can answer all these questions and provide all these items, then the job has a significantly higher probability of going well. Planning is the identification and scheduling is the handling of all these items

VALIDATION

One of the major pain points of modern CMMS is the amount of garbage that gets into the backlog and can't get out. The process of cleaning up the backlog is called validation. The planner generally does validation.

How to Validate the Backlog?

- Eliminate jobs already completed
- Eliminate duplicates
- Is the job a small job part of a larger job?
- Remove old jobs never to be done
- Clean up the work requested wording (be sure it is asking for what is needed)

Your role as the supervisor is to make sure every job properly is completed and is appropriately closed out on the CMMS. You also should be cleaning up the work requested wording and communicate any mistakes, contradictions, and missing information to the planner.

Determine the Ready Backlog

One of the jobs of scheduling is to release the maintenance work order to the ready backlog when everything needed is accounted for. The ready backlog is a list of which jobs have everything and are ready to schedule.

The ready backlog, when accurate, just by itself, will make your team more productive. Creating and maintaining your ready backlog is the single biggest thing to help productivity as well as morale.

Development of Work Programs

A critical job of the scheduler is determining the actual hours available for maintenance work next week. It should subtract all indirect time and divide the available time into buckets for PM, emergency, and backlog relief.

Your role in the work program is to communicate with the scheduler any indirect time or assignments. Some of the known losses would include:

Annual leave
Clean-up time (end of the day)
Disciplinary suspension
Holiday
Human resources requirements
Light duty
Meeting (other)
Meeting (toolbox, safety)
Meeting (RCA, quality, Kaizen)
Short term disability or sickness

Temporary assignment/relief supervisor
Training
Union business
5S activity

COORDINATION

Communication with operations and others is vital in a well-run company. The coordination meeting allows all stakeholders present their priorities and, hopefully, come to an agreement for the good of the whole company. Representing the maintenance team, the supervisor is a key stakeholder.

The supervisor should present what happened, and if it is off schedule, why it happened.

AGENDA FOR THE WEEKLY COORDINATION MEETING

Questions about last week's maintenance activity answered by the scheduler or the supervisor

- How did we do against last week's schedule, and exploration of any variance?
- What jobs were not completed as scheduled, and why?
- Did maintenance fail to get to them?
- Did operations deny equipment access?
- Did unscheduled jobs break into the schedule, and were they necessary?

Such questions highlight the underlying reasons for poor schedule compliance.

Questions about next week's work: The supervisor is a listener to this discussion between the scheduler and the production manager. If some jobs are impossible or unrealistic, the supervisor should chime in.

- The beginning of next week's schedule should be shared, showing resources available and demands already established (PMs, corrective work, carry-overs, project commitments, etc.)
- Next, the production schedule should be presented to clarify any operational support required from maintenance (setups, changeovers, change-outs of production expendables, etc.) as well as equipment access windows that can be utilized by Maintenance.
- At this point, all parties are ready for the give and take regarding who's job deserves higher priority in the interest of the overall operation. This continues until all available resources are committed in general terms. Some changes are inevitable when the scheduling process addresses the specifics.
- If participants continue to lobby for more resources, overtime and contractor support should be considered. This requires approval and funding.
- Finally, review all critical jobs that are delayed for lack of parts, engineering, approval, budget, or other reasons. Review these for possible expediting.

Productivity Increases

Some of the most significant sources of productivity increases:

- The workers have everything when they start work!
- Collisions between the requirements of labor, materials, tooling, and units to be serviced are far cheaper to resolve on paper, in advance, than on the shop floor.
- A reasonable amount of work is expected each day. Workers can be freed from a "hurry up" atmosphere one day and a "kill time" atmosphere the next. Given a reasonable amount of work, most mechanics will attempt to complete it each day.
- Emergencies are easier to handle when you know who is available and at what time. Most emergencies can wait for a short time. That small delay will buy you productivity by allowing the worker to complete the previous job.
- The rule of management control is to control the whole, control the parts. You must manage at the time and at the level where you can have the most beneficial impact.
- Uncover the many hidden operations problems. The solution to these problems (most are small) will increase productivity.

SCHEDULING UNCOVERS PROBLEMS IN YOUR OPERATION

Once installed, the schedule highlights areas where mechanics cannot do their job due to a problem outside their control and other issues with the old way of doing business. These previously hidden (or unpublicized) problems suddenly come into the foreground:

- Mechanics often get pulled off to work on non-productive activities.
- Stock room contributes to off-schedule conditions regularly.
- Failure to put equipment back into service when promised.
- Inability to get control of units when needed for schedule.
- Lack of cross-training causes clashes of resources.
- Failure to meet standards and immediately blame the schedule.
- Inability to handle emergencies, excess absenteeism, and unusual peak load conditions.

Section III

Supervisors' Toolbox

20 Transition from Worker to Supervisor

New supervisors often have trouble recognizing that the skills that got them promoted to supervisors are not the same skills that make a great supervisor. The greatest electrician in the world (the kind of person who can tell a wire gauge from across the room and know its insulation and ampacity rating) might fall on their face as a supervisor until they face the facts:

1. Skills need to change from technical to people/management skills.
2. The speed that you can do the job is unimportant. Your task now is to get others to do the job in a reasonable time. Berating people because they can't work as fast as you could is not useful.
3. The satisfaction and pride you used to get from doing a job right are gone. Your satisfaction will have to come from more abstract things like developing your people.
4. As people advance in the organization, they are expected to adopt a more long-term view of their job. You must shift from being repair-oriented to operation-oriented (looking at the whole operation for the longer term).
5. Your essential resources shift from the capabilities of your own hands and back (put into gear by your brain) to the skills of your subordinates (also put into gear by your and their minds). You will be measured by what you can control, not by what you can do.

The supervisor sandwich:

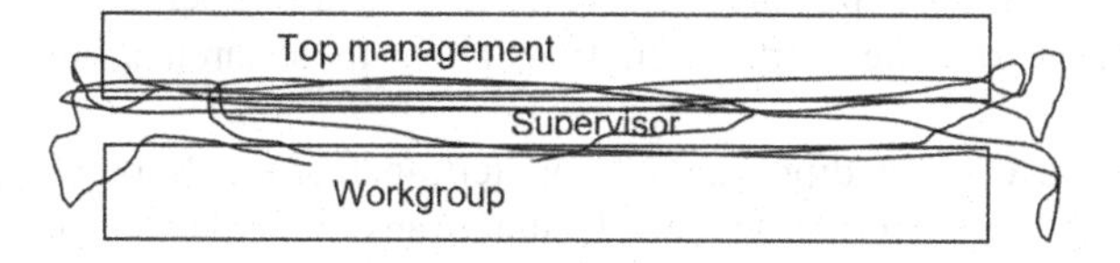

FIGURE 20.1 Supervisor sandwich.

SUPERVISING FRIENDS

Congratulations, you have been promoted to a supervisor! Oh dear, you now must supervise your old friends. This promotion may be an uncomfortable situation. There might be some left-over feelings about you being picked instead of them. You might even feel guilty.

The first rule is that, if the situation makes you uncomfortable, tell people the truth. Be honest if you feel uncertainty about the form of future relationships. Dr. Lee Minor says, repeatedly, when training supervisors, "If you can talk about an issue, you have the strength to deal with it." It is okay, even encouraged, to discuss your discomfort, guilt, or other feelings with your friends.

Beware:

- No matter what your relationship is on the outside, you have the right to be the boss on the job.
- Don't make your friend's into "boss's pets."
- Don't go overboard the other way either by dumping on your friends whenever anything goes wrong.

The trick is to be even-tempered and treat everyone fairly.

Expect yourself to have feelings about issues that concern your friends. It is normal to have these feelings. You still must act consistently with the best interests of your organization. If you cannot do this, you should transfer them someplace where they will be treated equally with everyone else (or ask for a transfer for yourself). If you do not do this, you are merely putting off trouble.

The same thing goes with supervising enemies: treat them the same as everyone else

SUPERVISING YOUNG WORKERS

One of the barriers to success with young workers is called transference. Transference is a psychological concept developed by Freud in the early twentieth century. It is a powerfully, simple concept. They will react to you, the supervisor, as if you were their parents, transferring all their emotions about their parents, onto you, without even realizing it.

There is a reverse condition called countertransference. Supervisors can suffer from this is especially true for parents. Countertransference means that you treat the young worker as if they were your children (so you assume they are a certain way). You may respond to them as if they had the same needs, habits, patterns as your children did. If you have no children, you may react to them as if they were your own younger siblings.

Because of transference and countertransference, the young worker and the supervisor are often not responding to the person who is standing in front of them. They react to their internal projections of each other, instead, in stereotypical (nonauthentic) ways.

Example: Tom, the new supervisor, is struggling to come up with next month's workforce assignment. Andy, the fresh-faced kid who has just been there six months, comes in to report on a suspicious bearing noise.

Tom, whose own younger brother always seemed to be whining about something, responds irritably, "Do you always have to come complaining to me? Can't you try to figure it out yourself?"

Andy, who was merely reporting a potential problem as he was instructed initially, but whose own father often burst into drunken rages, cowered his way out of sight.

Ideas:

- Exercise restraint with their youthful (sometimes misdirected) enthusiasm. Sometimes an analogy or a sympathetic word will redirect their energy in a more useful direction.
- Harsh treatment may create an angry, rebellious response.
- Wide swings in emotional states can be healthy. Exercise restraint and always exercise firmness. You allow the swings, and you require the job to be done.
- Peer pressure is one of the most potent forces of youth. Make them part of your team (even if they are doing menial tasks).
- Review the rules of behavior in direct language. Make sure they understand what is allowed and not allowed.
- It is ultimately your responsibility to ensure they do not get over their heads physically, emotionally, or in job skills.
- Since younger workers tend to have shorter attention spans, you might structure the job to change activities regularly.
- Demographics will make the young, entry-level employee more and more scarce. Your organization will directly benefit from a good reputation among your younger employee's peer group.

SUPERVISING OLDER EMPLOYEES

Supervisors must deal with two everyday situations. The first situation is the supervisor, who was promoted over older employees. The second situation is the hiring of older workers and supervising them for effective performance.

The first situation can be painful for all involved. The older employees generally have more significant experience, tenure, and may have many skills. Someone thought that they might have made a better supervisor (even if they did not want the job). Hurt feelings are common in this situation.

You were chosen by management. You have the right to be the boss. Their skills on the job may be impressive, but the skills needed from a supervisor are different than those required to do the job. The most effective method of working with anyone who has more experience in an area is to turn him or her into a resource. You are not trying to compete with them; they may know more than you. You are trying to enlist them and improve the overall effectiveness of your department or workgroup.

- Try to involve them in decisions.
- Ask them to be trainers when new people enter the group.
- Let them be advisors: to you, to any team efforts, or when management asks for input.

The second situation recognizes a tremendous resource that we have in our population. Our older citizens make excellent workers. Some generalizations, which are useful but not necessarily correct all the time:

- They are usually more reliable, on time, less frequently absent.
- They have a high standard of ethics.
- They are not usually as impatient for promotion or concerned about wage rates.
- They may be willing to work part-time (frequently they prefer it).
- They set an excellent example for younger workers and can be a stabilizing influence.

21 Building and Using Teams

Supervision is a team leadership role. Making sure that the team is taken care of and is operating healthily is part of the supervisor's job. Even when you are not supported by management, you are expected to manage your workgroup as a team (preferably as a winning team).

Teams: A team is a group of people linked in a common purpose.

The French language has a wonderful phrase for teamwork: esprit de corps. The spirit of a group that makes the members want to succeed. There is a sense of unity, enthusiasm, shared interests, and responsibilities.

The Happy Manager (http://the-happy-manager.com/articles/define-teamwork/) defines teamwork in a series of attributes:

- Trust in colleagues to deliver what they promise
- Willingness to help when needed
- Sharing of a shared vision of the future
- Co-operation and blending of each other's strengths
- Positive attitudes and providing support and encouragement
- Active listening
- All members pulling their weight and in the same direction
- Giving the benefit of the doubt
- Consensus building
- Effective conflict resolution
- Open communication

Take some care with teams

Be careful because the word "team" falls into and out of favor. Many organizations use the word team to mean any employee (some organizations use the phrase team member to help people feel part of something) or member of a department (like the sales team).

What is it about teams that makes them so well suited to so many of the issues of reliability?

Teams participate in RCA, PM improvement, PMO, RCM, Kaizen, and 5S. There are sound reasons for this. The first primary reason is that no one has enough experience, a wide enough point of view, or enough training to see all sides of an issue.

For our purposes, teams serve several purposes. Teams are the primary as a tool for remembering and reminding people of the activities needed to achieve their goal. Other vital purposes including getting resources, help when a member's orbit gets stuck when a member feels down and defeated when a member needs technical or other help, and a place to exchange ideas, tips, and techniques.

Tips for teams

- Be a team player. Respect each other's ideas. Question and participate.
- Relax. Be yourself. Be honest.
- Be willing to make mistakes or have an unpopular opinion.
- Realize and take advantage of Leverage: Huge opportunity for improved effectiveness because more effective teams improve all the participant's impact.
- Accept personal responsibility for team outcomes.

Structure for the fulfillment of our goals and projects: We use teams to help keep people in action. They remind people of the reliability leadership projects they want to accomplish. One person might forget. But a group will rarely forget.

Power of contribution: Workers want to make contributions to their friends and organizations. This is a case where the whole is greater than the sum of the parts. Much more power can be derived from teams than its individuals acting as individuals. The unit can become the check and balance for attitudes. Team members report results and bring problems to the team. The team can see if the operator is acting consistently with the philosophy.

Problem solving: One of the most important activities of the team is problem-solving. It is entirely appropriate for shop floor maintenance and operators to work on problem-solving projects. The most effective of these teams have a diversity of members. Your system should encourage ad hoc teams to solve problems.

22 Real Issues

Supervision is a slice of real life. The supervisor will eventually face most of the problems of society within their crew and company. While your judgment might be questioned, each crisis you face will build your experience and (hopefully) confidence.

ORIENTATION OF NEW EMPLOYEES

The best organizations consistently treat their employees as talented people. Just as you wouldn't invite someone to your house without introducing them around, you should introduce new employees to your company.

Their first few days are a time of high anxiety and high excitement. The introductions are also the time that your new employees form those early, crucial impressions of the people, the facility, and the work environment. They never get a second chance to create a first impression!

You are a busy supervisor. You might not have the time to spend because of the problems you face. Keep in mind that you hired the person to do a job that needs to be done, and that job is an important one. Time spent now will prevent problems in the future.

- One strategy is to spend some time setting up a visit schedule. Arrange for the new person to spend half a day just helping out, or shadowing each of several key people such as the planner, buyer, storeroom people, computer administrator, clerk, or other trade groups.
- Take these people around and introduce them to the entire crew. Make sure you say something that is both nice and true (this is John White, he's our best pneumatic person) about each person as you introduce them. Explain the new person's job when you introduce them (John this is Sarah Gleason, she's going to be responsible for our PLC diagnosis and repairs).
- Introduce them to support staff they might have to interact with during their day (purchasing agent, stores people, payroll/accounting people). Introduce them to their production counterparts. Follow the same rules as above.
- Introduce them to your boss. Tell your boss, in front of the person, your goals for the job and you feel that after interviewing 10 (whatever) candidates, this was the best person for the job.
- Tell the new guy your expectations about training, productivity levels, and your level of commitment to quality. Most people like to work for a supervisor with high (yet fair) standards.
- If they received the employee handbook from personnel, give them time to read it, and ask any questions. Answer the usual questions about when they will get their first paycheck, the status of vacation and sick pay, any probation periods, and all the general details (like the location of bathrooms!).

- Establish short- and medium-term training objectives. Set up the training process from the very beginning. Check back regularly.
- Everyone should go through elements of orientation every other year.

In an article by Dave Bertolini (July 2010) titled **New-Employee Orientation: Eight Elements of Success**, he discusses the kinds of materials you should gather (mainly if they are not supplied by HR)

Print out the companies' vision statement, mission statement, and shared beliefs or core values. Typically located on the company web page, these statements should be the first item in a new employee package. They describe the organization's reason for being. They also should be used to help make decisions. The supervisor should discuss them with the new employee.

Job descriptions: Detailed job descriptions are necessary to communicate what job the person is expected to do. These requirements should be clear and concise, with specific details. Each employee should have a copy of their position description and that of their supervisors.

Workflows: Workflows represent the most efficient and effective way to accomplish activities while affording opportunities for continuous improvement safely. The new employee should be shown all the inputs and outputs expected from their position in the workflow.

Responsibilities and accountabilities: Next, you should identify each step (in the workflow) by role or roles. You should review workflows with the employee to ensure nothing falls through the cracks and to reinforce responsibilities and accountabilities.

Organizational reporting and structure: You should review with each employee an updated organizational chart with reporting structures.

KPI (key performance indicators): Managers should review KPIs with new hires with special attention on the way the new employee can impact the KPI.

SUBSTANCE ABUSE

Did you know that alcohol is the leading health problem in the USA and that one in three hospital admissions is related to, caused by, or exacerbated by alcohol? It's not surprising that this social lubricant is also the leading cause of employee problems.

As supervisors, you are more likely to encounter alcohol as a personnel problem than almost any other. As a human being in the USA, you are likely to be affected by alcoholism. Everyone either knows someone with the issue or those of an employee, a loved one, or even a driver on the street.

Substance abuse is a tricky subject that we must discuss directly without flinching. Some supervisors have, or have had, problems with alcohol. Other supervisors may have alcohol problems at home or among their close friends and family. One of the most powerful indications of alcoholism is denial. Beware if you find yourself denying that someone around you has a problem (or you have a problem). Acceptance is the first step in dealing with the problem.

Drug abuse is related psychologically to alcoholism. Most, if not all, of the statements about alcohol are correct about other addictive drugs as well. As of this writing, marijuana is legal for recreational use in several states in the US and all of Canada.

Since the laws have changed, some organizations have not caught up in changing their policies (you may be breaking a company rule even though you are not committing a crime). Get clear guidance of where your organization stands on this issue, hopefully, before it comes up.

Warning signs: Many people drink occasionally and have occasional problems. These warning signs are geared toward people who change for the worse and stay that way.

- Be alert for changes in the presentation of the person (personal grooming, hangovers, speech patterns).
- Changes in productivity: Is the person who could rewire a control circuit in their sleep now making stupid mistakes regularly?
- Changes in attitude: Excessive worry, moodiness, or over-reacting to situations.
- One symptom is more emergency phone calls, more absenteeism (employee's spouse or child could be a problem, and the employee has to cover for them).
- If you ask, "Is there a problem," you will usually face a wall of denial.

Consider your workforce. Is there anyone that you suspect might have an alcohol or drug problem? Always consider that the problem might be related to legal prescription drugs, over the counter drugs or medical conditions. That is why if you determine there is something wrong with the person's performance, demeanor, or inappropriate attitude, you should bring in your HR department to identify the next steps.

What you, the supervisor, can do about alcohol or drug problems:

- Most professionals agree that substance abuse is challenging to treat. Most advise supervisors to leave any counseling to the pros.
- First, try to talk to the person, especially if you have a good relationship.
- Treat infractions to the rules the same as any other infraction. Give consistent, firm, fair discipline. Do not allow the alcoholic to "get away with" anything.
- If a safety problem exists for the person or anyone in the crew, immediately remove them from the situation. You can deal with the paperwork later, if necessary.
- Consult resources inside your organization. Your organization may have a policy for dealing with these types of problems. You can also consult with your boss or an older supervisor for advice.
- Resist the temptation to cover for the person for too long or you will become what is called an enabler and part of the problem rather than part of the solution. Caution: People with alcohol problems in their family history sometimes feel a need to protect their subordinates with alcohol problems.
- Your organization might have an EAP (Employee Assistance Program) that will help fund professional support. You can provide support to an employee getting help.

- If your company doesn't have any policy or an EAP and you feel as if you are on your own, you must consider the consequences of anything that you do even if you avoid the issue. You are letting someone work while under the influence is dangerous to your crew, the person, and to anyone around them.
- Recommend AA: there are chapters in all cities, most towns, and some companies even have in-house chapters. If the problem is an alcoholic spouse, recommend Al-Anon. If the person has alcoholism in their family history, locate an ACOA (Adult Children of Alcoholics) chapter through AA. Drug abusers should be directed to NA (Narcotics Anonymous).
- You can't save them or do the hard work of recovery for them. You can be supportive of their recovery process if they are willing to go that route.

Thanks to Marjory Levitt, Ph.D. (my sister), a licensed psychologist and consultant on alcoholism, to the Philadelphia court system, retired Professor of Psychology for material in this section.

DISCIPLINE

You carry out company policy. If other methods don't work, you will be required to use discipline. As a supervisor, you are expected to deliver the discipline fairly and consistently. The purpose of discipline is to correct a behavior, not change a personality, or punish the person. Effective discipline changes the practice in such a way that we don't have to do it again and again.

Few people are surprised that the rules of good, effective discipline are the same at home as in the office (except for perhaps the business concept of termination!). There are good psychological reasons for the similarity. People learn about discipline and punishment at home as children and naturally respond in similar ways.

Many researchers have done excellent work in this area. Some of the material in this section is derived from the excellent work of Dr. Walt Lacy, W.H. Weiss, and James Van Fleet.

Three Rs of Discipline

You must have a valid reason for any form of discipline. The infraction must break a written rule. Finally, the punishment must be relevant to the offense.

Reason

To establish the reason, you must collect relevant facts. Frequently, the events will indicate that no problem exists. A discussion with the person (important—do this without any accusations) is often useful to collect data. Several reasons are considered appropriate for discipline.

- Illegal acts such as illicit drug use, theft (consider a value limit), sabotage, or industrial espionage. Does your company have a policy for involving the police when laws are broken on company property or time?

- Unsafe acts involve intentional behavior that could lead to an accident. In this case, deliberate means that the hazardous act (like horseplay around machines) was on purpose. Unintentional, unsafe acts are usually caused for more training, not discipline.
- Insubordination is the deliberate refusal to follow direct instructions from a supervisor (where the instructions don't endanger the person, the supervisor has jurisdiction, and are otherwise legal). Insubordination includes disobeying long-established rules.
- Insubordination has many different styles, including overt, covert, direct, and indirect. Frequently, you must peel away the veneer, look below the surface, and see what the behavior is. One common type of insubordination is called passive-aggressive. In this type, the person might agree to your face while doing what they want behind your back.
- Continued shoddy work is the most challenging area for discipline because it may involve subjective judgments and opinions. There are three reasons that people continue to do shoddy work.
  1. They don't know the job. This is a problem because it is widespread, and both supervisors and workers don't like to admit it. The responsibility to convert a "don't know" into a "know" is the responsibility of the supervisor. Even a skilled mechanic will sometimes run into problems they don't know how to solve and don't want to admit it.
  2. They don't have the aptitude for the job. They might not be strong enough, smart enough, or have the right-thinking style. Consider a maintenance draftsperson who can't visualize. They are disabled in that field. If the same person were an estimator or planner, their inability to visualize wouldn't be as much of a handicap. The responsibility for an aptitude problem is also the supervisors.
  3. They have an attitude problem. We finally have a reason that is not the supervisor's fault. Yet we frequently create this type of difficulty out of the first two. Consider the surprising fact that 92% of poor performance is related to reasons 1 and 2 above.

Rule

For discipline to be effective, the person must break a written or a well-established rule. For maximally effective discipline, the person had to know ahead of time both the rule and the consequences of breaking it.

If your organization does not have written rules of conduct, you might consider asking if you can organize a group from different organizational levels to create one.

Idea: When we put in our first employee handbook, we included a space for the employee to sign for receiving the book and a second place to sign after they had a chance to read it.

Relevance

The discipline must fit the situation. Effective punishment should be uniform, corrective, and progressive. Similar offenses should receive the same punishment. Always give the person a way to right the wrong (fixing a lousy job, apologizing, etc.).

This allows the person to correct the problem they caused. Punishments should be progressive. The first offense might draw a light reprimand, while the second a written reprimand and the third, time off without pay.

Checklist of considerations:

___ How serious was the rule infraction?
___ What were the circumstances?
___ How long has the person been employed by the company?
___ How was the person's past conduct?
___ When was the last time the person received discipline?
___ What is the usual discipline for this infraction?

Progressive discipline (outline the consequences of a further breaking of the rule and include that in your notes)

1. Verbal comment (no notation in the file)
2. Verbal warning (notation in file with date and comments)
3. Verbal warning (in front of a witness, notation in the file, etc.)
4. Written warning (copy to file)
5. Written warning (employee's signature on notice showing they received it)
6. One day suspension with pay (note to file, written assignment to the employee: Is this job important to you, how can we remedy this situation? Some organizations call this a decision-making leave.)
7. Suspension without pay (note to file)
8. Termination

Most organizations establish some period to clear a person's record of a level of disciplinary action. A written warning for lateness might be removed in a year, while a one-week suspension for insubordination might stay on the record for two years or more.

TERMINATION

You have tried everything, and you have decided that your organization cannot afford to keep this person. You have discussed the decision with your supervisor (supporting your case with the written records). You know you need to terminate this person.

The first problem is a fundamental unwillingness to make the decision and stick to it. We all want to be "good guys," and any termination can make us look bad.

How to Terminate Someone (Legally)*?

1. Make the decision and stick to it. Consult an employment contract, bargaining agreements, and all informal agreements. Ask if you are terminating this person within all the agreement provisions.

2. Pick a termination date. Determine benefits such as sick pay, vacation, pension, comp time, health plan continuation, etc.
3. Look at your organization's procedures for termination. Is your written case built consistently with your processes?
4. Consider the critical legal issue of unequal treatment. It is illegal to treat certain groups differently. Certain groups, such as women, minorities, the disabled, sometimes age, religion, sexual preference, are categories protected from unfair treatment by law. You must be sure (you may have to defend yourself on this) that the person is not being terminated for any reason not directly related to the job. Be careful about issues of strength or endurance (that these attributes are needed for the job and not a way to exclude women or older people).
5. Look at your written record and verify that all steps taken are documented, and all appropriate forms were signed. Play the devil's advocate and see if your original case is entirely defensible. Have you protected your organization's position from a lawsuit at some later date?

VOLUNTARY TERMINATION (PEOPLE WHO QUIT)

The founder of one of the leading temporary agencies for accounting and data processing, Robert Half, says that a resignation should never come as a surprise to the immediate supervisor. He says that there are several common symptoms to watch out for.

(Note that some of these symptoms can be evidence of a variety of problems or changes to a person's life. We suggest that you be on the alert. You might be able to head off an unwelcome termination or help with some other problem.)

Ten Signs That You Might Be Facing a Resignation

1. A sharp increase in personal use of the phone or cell.
2. Noticeable improvement in grooming (may be interviewing).
3. Longer lunch hours, increase in personal business time, increase in absenteeism.
4. Less communication with management (wants to be less visible).
5. Changed vacation patterns (to follow-up on job leads or take a vacation before they start a new job).
6. No longer stays late or takes work home (change in attitude toward company).
7. Neater work area (person removing personal items).
8. Quality of work and commitment to completing jobs changes.
9. Disregard for the system (don't care attitude).
10. They will not take on any long-range projects.

Issues with Downsizing

- Can you capture the knowledge going out of the door?
- Build detailed job files to pass on information.
- You know how hard it is to replace someone on the floor and how long it takes so keep in touch with the best ones.
- Perhaps there is a part-time situation.

NOTE

* **Consult with your organization's personnel department or legal counsel for specifics on what organization rules and laws apply in termination in your area.**

23 Quality

Maintenance work is subject to the same quality issues as production work and a few unique contributors as well. In maintenance, there can be quality defects in craftsmanship and materials. Also, there are potential issues in identifying and then troubleshooting the original problem (doing an excellent job to the wrong equipment). The nexus of quality is the supervisor.

W. Edwards Deming transformed the Japanese industry using statistics and by using quality to drive all decisions.

Deming was considered the quality guru for the last generation of Japanese quality experts. The quality award in Japan today is called the Deming Award. He had much to say about quality in manufacturing. It should be no surprise is that Deming's points apply to maintenance also.

He proved that lower costs come from an ongoing commitment to quality. The question is how to apply his principles for transforming the business into a maintenance department that may not have the support of the rest of the organization.

The perceived quality for maintenance work often stems from the need for, or the consequences of the repair, not the repair work itself. The emotional context of the response is also tied up in perceived maintenance quality as well; a surly, dirty maintenance technician seems to be low quality even if their work is superb.

In some circumstances:		
maintenance quality might	=	no downtime
In others:		
maintenance quality	=	no scrap waste produced
maintenance quality	=	fast start-up
maintenance quality	=	safe operation
maintenance quality	=	on-time delivery
maintenance quality	=	lowest cost
maintenance quality	=	quick response
maintenance quality	=	no calls from the public
maintenance quality	=	a comfortable building
maintenance quality	=	safe indoor air quality
maintenance quality	=	no repeat-repairs
maintenance quality	=	keeping the unit in spec
maintenance quality	=	no interruptions
maintenance quality	=	a satisfied user

Every maintenance operation should define quality in the way most useful to their operating environment.

We know much of what is needed to produce quality production and a safe work environment. What is required is to review the list of all the elements of a planned job.

This list presupposes the most crucial element of quality has been handled: that the worker has complete knowledge of the scope of work and the competency to carry it out.

1. Person(s) with the right skill(s) to perform the job, who is physically and mentally able. The key here is that the right person is chosen for the job, and they are ready to work to their ability (not sick, not preoccupied).
2. Correct parts, materials, supplies, consumables are available for the job. When the appropriate components are not available, the tradesperson must improvise. While improvisation is great in a theater, it introduces potential quality problems in a maintenance situation.
3. Correct special tools produce higher quality work. Frequently, the difference between a professional and an amateur is their tools. The wrong tool will also require the tradesperson to improvise.
4. Adequate equipment for lifting, bending, drilling, welding, etc. Once again, the difference between a professional and an amateur is their equipment. The wrong equipment will also require the tradesperson to improvise.
5. Safe access to assets (decontaminated, cooled down, etc.), secure work platforms, and humane working conditions. Working conditions are essential for quality work performance. Areas that are too hot or cold, areas where there are not safe work platforms or reasonable access to the assets put the workers at a disadvantage in performing reliable work; all these will compromise quality.
6. Up-to-date drawings, wiring diagrams, and other information can be the single most crucial element since most complicated jobs should not proceed without drawings or wiring diagrams. The lack of good drawings can introduce all sorts of problems.

These items were maintenance-specific items (unique to successful maintenance).

In 1950, WE Deming went to Japan and talked about what gets in the way of quality in general. These diseases and obstacles promote defects in quality **BUT also in SAFETY!**

DEMING'S DEADLY DISEASES

- Lack of constancy of purpose.
- Emphasis on short-term profits.
- Evaluation of performance, merit ratings, or annual reviews.
- Mobility of managers and job-hopping.
- Management use only visible figures, with little or no consideration for data that are unknown or unknowable.

Obstacles to Improvement

- Hope for instant pudding.
- The supposition that solving problems, automation, gadgets, and new machinery will transform the industry.

- "Our problems are different." And the search for examples.
- Obsolescence in schools.
- Poor teaching of statistical methods in the industry.
- "Our quality control department (safety department) takes care of all of our problems of quality."
- "Our trouble lies entirely within the workforce."
- False starts.
- "We installed quality control (a safety program)."
- The unmanned computer.
- The supposition that it is only necessary to meet specifications.
- The fallacy of zero defects.
- Inadequate testing of prototypes.
- "Anyone that comes to try to help us must understand all about our business."

IATROGENIC FAILURE

When you fix things, sometimes you make them worse or worse, break something working fine before you got involved. Iatrogenic: Referring to injuries caused by a doctor or surgeon or by medical treatment or diagnostic procedures (1924).

Every time you touch a machine (for any reason), you run the risk of messing it up!

That risk goes up as you interact with the machine more intrusively.

Ever hear the complaint, "After you guys PM my machine, it NEVER works right?" IIoT great news: The sensors are doing the inspection non-interruptedly so less opportunity for iatrogenic failures.

24 Supervisor Time Management

You are leveraged. That means that if you can improve your productivity, you leverage your team toward enhanced productivity. Time management is a life skill and will be useful to you during your whole career. Understanding the dynamics of time management will also reduce your stress levels.

THE NATURE OF TIME

Time is the only truly nonrenewable resource. Unlike energy, it cannot be saved or created. It is also the single resource that everyone has the same amount of. A leader of a multi-billion-dollar organization has the same number of seconds and hours in the year (31,557,600 and 8766, respectively) as you do. In the amount of time, we are truly equal. What you do with what you were dealt with is a game worth playing.

TIME MANAGEMENT = SELF DISCIPLINE

Time management is a life skill. A life skill is a skill that the supervisor should hone and sharpen throughout their entire life. However, high you rise in your organization, time management will make you more productive. Good time managers have time (and, sometimes, more important, energy) for the fun things in life like family, hobbies, and wasting time.

RIGHT MIND, RIGHT THING, RIGHT TOOL

- Right Mind
- Right Thing
- Right Tool

Time Management Has Three Components

Three components regulate the effective use of your time. While you can get things done with one or two or without alignment of the three, effectiveness and efficiency come from all three being actively used.

The Right Thing

It could be the most important. If you are doing the right activities, you will eventually reach your goals or get the most important things done. In this sense, it is essential to align the activities to your mission (or your organization's mission).

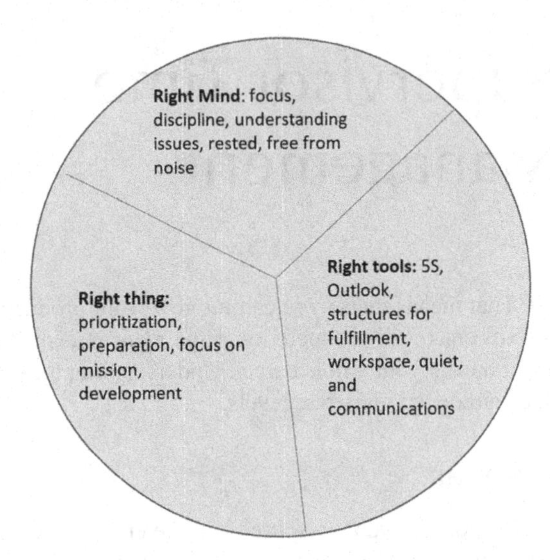

FIGURE 24.1 Complete approach to time management.

The "Right thing" also presupposes that of all the activities you face, this one now is the one that will "move the needle" the most toward your goals. The other aspect is to be prepared when you do something; you have done your homework.

The Right Tools

Having the right tools allows you to leverage yourself and get the most return from your invested time. Tools (at least high-tech tools) seem to be getting better all the time. Between our Apps, Google, tablets, and our smartphones, we can choose to use our time intensively. Of course, we see some downsides to this, but we can deal with the pitfalls later in the discussion.

The most crucial function of most of the tools is called a structure for fulfillment. This idea (described more fully later) supports you in delivering on your promises. Someone who delivers on their promises is someone people want to work with!

The environment is crucial. Simple things like having what you need, limiting distractions, and quiet; facilitate good work done on time. Of course, if you work best in the middle of YouTube videos, music streaming, and chatting, then have at it.

The Right Mind

Are you in a conducive mood in terms of sleep, hunger, free from distractions, and as healthy as you can be (given your circumstances)? Do you have the energy to engage your chosen assignment?

> "The ability to concentrate and to use your time well is everything."
>
> Lee Iacocca

WHAT ON EARTH IS THIS "TIME" WE ARE SO HELL-BENT ON MANAGING?

Oxford dictionary makes it clear, "The indefinite continued progress of existence and events in the past, present, and future regarded as a whole." They go on to clarify: "the continued progress of existence as affecting people and things."

That doesn't help. We know there is clock time. Clock time is defined as "time or an amount of time as reckoned by a conventional standard." But that too doesn't seem useful either.

The second used to be defined as the fraction 1/86,400 of the mean solar day. That caused some problems since the Earth wobbles a bit. It took until 1967 to agree on the definition of a second (According to NIST standards: The second is the duration of 9 192 631 770 periods of the radiation corresponding to the transition between the two hyperfine levels of the ground state of the cesium 133 atoms.). Again, this is not too useful (unless you are a physicist or astronomer).

The problem we have with time can be answered by answering the questions:

What is time? Is time a thing? Is time an idea?

The answer is YES!

Time is a thing when you are thinking of seconds, minutes, and hours. Clock time is a thing. Time is an idea when you try to manage it. We will call time managed brain time.

Brain time is a slippery concept. Did you ever notice how much longer an hour waiting for a dentist is then an hour when you meet the love of your life? How can this be? It's brain time.

I was on a flight to Chicago to connect on a trip to Portland, Oregon. We left late, and it looked like I would miss my connection. I got word that my outgoing flight was a little late, so I had a chance to make it. We landed, taxied, and pulled up to the gate. We disembarked. The time taxiing seemed to be the longest I could remember. It also took forever to empty the plane. Each minute I stood waiting seemed like an hour. Eventually, I missed the connection. The feeling of the speed of time changing to suit the circumstance is brain time.

Our brain can make it seem that we are overworked, rushing around chronically behind, or it can slow reality down, such as when you hit the perfect curve going down a ski slope (and time seems to slow down). We are pretty sure the atomic clock in Boulder Colorado doesn't speed up or slow down to suit our mood or activity. So, something else is at play.

So, in brain time, your job can be a pressure cooker. Emergencies, short staffing, vendor problems, high customer expectations all contribute to being stressors on the job. While removing the stressors is impossible, changing your attitude might be possible.

When we talk about time in this book, we are dealing almost exclusively dealing with brain time. A person who feels mastery over his/her environment transforms the 'bad' stress into pleasant excitement (transformation of the relation of the person to their own brain time).

Working on your time management is a life skill that helps people feel mastery. The goal is not to feel in control because control is an illusion. The purpose of time management feels that you can master anything that comes at you. Mastery of this type pays dividends in effectiveness, well-being, potency, and even success.

It starts with knowing how you spend your day. When you take the time to write it out, you will probably be surprised.

CHECKLISTS

One simple improvement saves thousands of lives and billions of dollars.

In settings where the outcome is critical (aviation, power plants, medicine, or military operations), checklists are widely used.

What we might not know is that the use of checklists has been proven useful in widely different situations. In as different environments as production floors, daycare centers, and political campaigns, checklists play a prominent role.

The idea is simple: Design a checklist for any repetitive activity, distribute, post, and use it. Here are a few tried and true simple rules to make it work best.

- The list should be on a single side of the paper. Use a large font that is easy to read.
- The most effective checklists are short and quickly completed. Many can be run through in under a minute.
- They would typically have 7–10 or so items on them.
- Post them and use them!

A checklist is simply a reminder system. It consists of things that you already know (hopefully) but might forget for a particular event. Whenever something new goes wrong, you can review it to see if it is likely to happen again.

The checklist keeps the practices that make a successful event (like a meeting) right there in front of you. That way, even if you are tired, have a headache, or are preoccupied, you won't miss something important.

Try checklists for some of these:

- Meeting prep
- Safety walk down
- Job scoping
- Job planning
- Scheduling

You can also use the checklist to remind yourself to try new ideas, techniques, or practices. For example, you could rotate items into the list that you would like to try out (such as some of the more exotic solutions or products)

The power of the checklist is in the execution. It only has a positive effect if we use it! There is a whole chapter on checklists for meetings in the book "10 Minutes a Week to Great Meetings."

Seven Daily Habits

We can thank Robert Allen and his associates at Challenge Systems for taking the work of Napoleon Hill (this is old school stuff, really old) and others to come up with the seven daily habits of success:

1. Spend some quiet time every day doing the following three activities.
   a. Making sure you have a coherent purpose in your work situation (can keep this private).
   b. Spend a few minutes each morning (or the night before) planning your day.
   c. Visualize your day ahead of time being the way you want it to be.
2. Your first task every day is to schedule and do the thing that you fear or resist most.
3. Schedule and complete the three most important tasks (called "bottom lines") to support your purpose in the following areas: Work, Others, Self.
4. High efficiency depends on a healthy body. Be sure to plan for adequate rest, proper nutrition, and daily exercise.
5. Write down any ideas that come to you about your situation at work or home.
6. Be a teacher or mentor to another person.
7. Review your day to see where you did well and where you would make changes.

Which habits do you already practice? how to tame time killers!

Ideas for Getting Control

- If you feel overwhelmed with too many projects, sit quietly for one or two minutes, and allow all the worries to surface. List the concerns that surface. The concerns that occur first might be stopping you from the rest of your work. If possible, try to put the worry list to bed (do them or schedule to do them) first.
- After handling the concerns, decide what your most important jobs are. Set priorities. Do your highest priority items when you have the best energy level.
- Gain efficiency by grouping related activities together. For example, make all telephone calls together or assign all estimates to WOs at the same time.
- Use scheduling concepts to gain control of your projects. Divide larger projects into sub-projects. This division is fundamental to the philosophy of scheduling. Give yourself the extra motivation of allowing a completion (of a sub-project) every day. Reverse-load larger projects. Reverse loading starts with the completion date and works backward to the beginning of the project. This gives you logical sub-projects and milestones to see if you are on schedule.
- Always work to complete what you start. Going back will cost you time. Experts even extend this to reading your E-mail. Never open an E-mail and close it to deal with later. Open it once and deal with it.

- Use a diplomatic means to end telephone conversations that aren't going anywhere. They include "Glad you called, I have a meeting, so can I call you back?" then call the person back at 4:50 pm (right before they go home) to "chat."
- Look at your junk E-mail. Question which types are useful. Ask your system administrator to block useless messages. You can unsubscribe from reputable firms' lists.
- Take control of parts of your day. Do not allow interruptions of specific tasks—instead, reroute calls to a crew member, clerk, or another supervisor. Use this time for your bottom lines and your 20% activities. The part of the day to focus on is your high energy time.
- Some meetings are an energy killer. Never schedule long meetings during your high energy times if you can help it. The best time for long meetings that you control is around lunchtime or at the end of the day. High energy meetings can be inserted during high energy times if they are short and aggressively functional.
- Homework. The way to get the best return on investment from meetings that you run is to insist that people do their homework. The least efficient meeting is one where people sit around and watch each other think because no one has bothered to prepare in advance! Always schedule prep time for meetings that you attend.
- Train yourself to be able to throw things away and put things away. You may walk around a pile of magazines for months before realizing they can be placed in the circular file. Look around your workspace once a month; throw some junk away. Do it until only current or essential stuff is left. (No doubt your whole maintenance shop area can benefit from this too. Look up 5S for ideas.)
- Put T.O. dates on all files. (T.O. stands for throw out.) These T.O. dates will keep your records clean. Every six months, review your documents and throw away the old, unnecessary ones.
- Get used to the fact that you don't have all the answers. When something stumps you, restate the problem or spend time trying to isolate the core of the problem. Seek out people with knowledge, inside and outside your firm. Learn techniques for root cause analysis to structure your problem-solving.
- Attack your overweight contact lists (in the old days—bulging Rolodex), on your SIM card and Outlook files. The business would screech to a halt without these phone number/business card organizers. With many good things, a little is great, but too much is not better. The goal is to have a slim Outlook Contacts file that is a delight to use. Limit it to your current, most often used numbers. Copy the PST file or Export the Contacts before you edit the list down. File the cards that are not current or important right now. Develop a habit of writing the date and source on every contact file.
- Buy speed-listening equipment. Get podcasts, spoken books, and listen to them.
- Note: The average brain can process information faster than most people can talk. Firms have developed speed listening capabilities (some podcast players allow speed to be set). These players speed up the content by skipping very short

segments and playing the rest of the tape at normal speeds. The pitch of the speech and music is typical, but the elapsed time is variable up to 1/2 of normal.
- Know and drill yourself on doing jobs to an appropriate level of quality. For example, a punch press tool designed for millions of pieces will be made to a different quality standard than a temporary means to make 15 pieces. Some of your projects need to be done 'quick and dirty,' and others need to be excellent. Know the difference. Inappropriate quality is a time killer.
- Open days for vendors and suppliers. No drop-in meetings without appointments except these times.
- Restrict your computer time (especially email and social media) to fixed intervals. Set up two screens for your computer so you can alternate activities when the system is slow.
- Consider delegation by where things are located, skills, materials, and other factors.

FOUR TIME-MANAGEMENT PROJECTS TO GET YOUR LIFE BACK!

New Way #1 According to Jeffrey Mayer in his book *If You Haven't Got Time To Do It Right, When Will You Find Time To Do It Over?* Your first task is to (ouch! Sorry!) clear off your desk. Project #1 tells you how to go about doing it:

New Way #2 Your time is precious. In "*The Ninety Minute Hour*," author Jay Levinson shows the value of time by telling us to get 90 minutes of work out of every hour. Project #2 explains how to supercharge your time.

New Way #3 No one with an active mind can do everything that they think needs to be done. This unfinished business is a source of frustration. Knowing that you can't do everything, you must focus on the critical few things that make a difference. Learn how the 80/20 rule works in your business life. Project #3 looks at your master to-do list and applies the 80/20 rule.

New Way #4 Your energy goes up and down all day long. Figure out when your energy is highest and lowest. Establish patterns of energy. Look into behavior that drains you of energy such as big lunches, certain kinds of meetings, or people. Leave high energy times for more top priority work, and jealously guard this time. Project #4 helps you plot your energy level.

Project #1 Clean Your Desk, Organize Your Office

A. Set aside a length of time when you won't be interrupted.
B. Get an entire box of manila folders. Put them in order with the tabs alternating left-center-right-left-center-right.
C. Get a blank piece of paper (full-sized) or open a document for your Master To-Do list. The list can also be Tasks in Outlook.
D. Put all your papers in a pile. Include the little half, quarter sheets, envelopes, napkins, and everything else you were going to get to.
E. Go through every piece of paper and be ruthless about throwing as much as possible away. Separate into a new pile, the documents you need to keep. If there is an action that needs to be taken, save the article in this new pile.

Now take a break for a few minutes and look at what you accomplished. All your old piles have been consolidated into a new smaller collection. Maybe half of your old stuff is now in your trash can. That's progress!

F. Back to the grindstone. Start with the top paper and ask yourself some simple questions:
   - One minute to 4F each paper. 4F: File, Finish, Forward, Flush
   - Is there any work that must be done with this? If so, add that assignment to your master to-do list.
   - Should I keep the piece of paper? Remember if you recorded the assignment. If it's a keeper and no file already exists, prepare a manila folder.
   - Keep this up until you reach the bottom of the pile. Do not do the work now unless it can be handled within one to two minutes.

G. Remove all office supplies such as tape, staplers, and paperclip collections from the surface of your desk (visual field). Attack any surface of your office that accumulates papers for more than a very short time (the paper turns over in 1 shift).

H. Consider this a 5S exercise and have a place for everything–everything in its place. Oh yes, get some Pledge, 409 or Fantastic and clean it too.

I. Apply the same standard to your files and drawers. Review the information and determine if you need the data, or if there is any action that you need to take. Clean your drawers of the debris that accumulated over the years.

J. You now have a clean desk, that is the base condition. You also have a complete master to-do list that could be up to several pages long. You will find that a clean desk encourages productive work habits. You put your work on your office, complete it to the level possible at that time, put it back into its proper files, and have a clean desk once again.

Every Friday (or periodically), return your work area to base condition.

Project #2 Ultra Efficiency. You Have Three Major Tasks

1. Get your immediate job done.
2. Allow time for the significant critical jobs/analysis.
3. Educate yourself—make yourself more valuable.
   A. Look at areas where you could do two things at once without sacrificing the quality (or safety) of either. One example would be your daily commute. An hour's commute could translate into over 200 hours a year of learning time using CDs and podcasts. Remember that 200 hours is equivalent to three college-level courses (homework time included). Additional Ideas: make use of time spent waiting for meetings, waiting for airplanes, driving your family around, exercising, taking a bath, mowing the lawn, etc.

   (My local library has over 600 CD sets for borrowers on various business topics!)

   B. Learn to use your voice memo app on your phone or carry a micro voice recorder to record ideas, notes, letters, and instructions. This is an excellent idea if you have the staff support to transcribe your tapes. Some software systems (such as Dragon Naturally Speaking) will attempt to transcribe your words into typed test files. The transcription should use a separate page for ideas on each project which can be filed in the project file or virtual folder. Your brain is your most powerful tool.

C. Use your crew to delegate some of your tasks. A well-trained team will multiply your effectiveness.

D. Learn to touch type and improve your use of the computer. Once you learn, you will type at least as fast as you write, and with less fatigue and greater accuracy. Some people have trouble learning to type. For those people, the computer comes to your rescue. Consider voice recognition software (such as Dragon Naturally speaking) http://shop.nuance.com/DRHM/servlet/ControllerServlet?Action=DisplayProductDetailsPage&SiteID=nuanceus&Locale=en_US&productID=111903500):

Learning to Type Web resources:

- http://www.customtyping.com/
- http://www.powertyping.com/
- http://visibone.com/type/

E. Use E-mail attachments of digital photographs, video to transmit ideas (particularly drawings) to remote sites, vendors, users. It's faster than mail and more accurate than verbal descriptions.

F. If you drive during the day as part of your job (managing several sites), use your Bluetooth to car interface or get a hands-free set up for your mobile phone (by law in several states). This way, the little ideas that occur to you can be transmitted directly into action. Always pull over when doing intensive talking on the road.

G. Learn to speed read.

If you think that it would be a waste of time, set up the stopwatch function on your phone and try this:

Start

Time your reading speed. Read this paragraph. Note the elapsed time. Divide the total word count (100) by the number of seconds it took and multiply by 60. A good fast pace is 400 words per minute with good comprehension. Speed readers top out at 1500 to 2000 words per minute. The funny thing is that people with faster reading also seem to have a better understanding and retain more. Many maintenance supervisors read at 100 or fewer words per minute. How much more could you cover if you read at 1000 wpm?

Reading rate words per minute = (Words/seconds) * 60

End

Speed reading references:

http://www.mindtools.com/speedrd.html
http://english.glendale.cc.ca.us/methods.html

Project #3 The 80/20 Rule

An Italian economist, Vilfredo Pareto (1848–1923), discussed an informal rule now called the Pareto principle or the 80/20 rule. It can be stated for most management situations, as 80% of the action comes from only 20% of the actors. We find that 80% of our employee problems come from 20% of our employees. Likewise, 80% of our emergencies come from less than 20% of the equipment. (His research stated that in Italy's economy in the nineteenth century, most of the wealth, maybe 80%, was controlled by a few families—sound familiar?)

80% of our results will flow from 20% of our activities. If we identify these critical few activities and increase our time commitment to them, we can double or triple our results each day. This is not merely because some critical activities might not be obvious. Do the best you can, and your choices will improve as you learn more.

Your exercise is to go through your master to-do list and examine each item. Divide the list items into the 20% important and highly leveraged activities and the 80% low leverage activities.

The 20% activities should be scheduled first, making use of your high energy intervals throughout the day (discussed later). The top three of these activities become your bottom lines. The bottom line is, "if you do nothing else today, you will at least do your bottom lines."

Use the low return-on-investment activities, the 80% activities, to fill in around the 20% activities.

Steven Covey's matrix of action:

	Important	Not Important
Urgent *Beware of the tyranny of the urgent!*	1. No problem for maintenance leadership. You're good at this, or you find another job.	2. A problem because we tend to jump when the phone rings. Work to manage this.
Not Urgent	3. It is an area of focus. Longer-term issues with no constituency now.	4. not a problem – learn to procrastinate with these.

Project #4 Energy

Establish your energy level by times of the day with a code you can make up

	Monday	Tuesday	Wednesday	Thursday	Friday
6:00 am					
6:30	______	______	______	______	______
7:00	______	______	______	______	______
7:30	______	______	______	______	______
8:00	______	______	______	______	______
8:30	______	______	______	______	______
9:00	______	______	______	______	______
9:30	______	______	______	______	______
10:00	______	______	______	______	______
10:30	______	______	______	______	______

11:00	______	______	______	______	______
11:30	______	______	______	______	______
12:00 pm	______	______	______	______	______
12:30	______	______	______	______	______
1:00	______	______	______	______	______
1:30	______	______	______	______	______
2:00	______	______	______	______	______
2:30	______	______	______	______	______
3:00	______	______	______	______	______
3:30	______	______	______	______	______
4:00	______	______	______	______	______
4:30	______	______	______	______	______
5:00	______	______	______	______	______
5:30	______	______	______	______	______
6:00	______	______	______	______	______
6:30	______	______	______	______	______
7:00	______	______	______	______	______
7:30	______	______	______	______	______
8:00	______	______	______	______	______
8:30	______	______	______	______	______
9:00	______	______	______	______	______
9:30	______	______	______	______	______
10:00	______	______	______	______	______
10:30	______	______	______	______	______
11:00	______	______	______	______	______
11:30	______	______	______	______	______
12:00 am	______	______	______	______	______
12:30	______	______	______	______	______
1:00	______	______	______	______	______
1:30	______	______	______	______	______
2:00	______	______	______	______	______
2:30	______	______	______	______	______
3:00	______	______	______	______	______
3:30	______	______	______	______	______
4:00	______	______	______	______	______
4:30	______	______	______	______	______
5:00	______	______	______	______	______
5:30	______	______	______	______	______

- Circadian rhythms are physical, mental, and behavioral changes that follow a roughly 24-hour cycle. These rhythms can change sleep-wake cycles, hormone release, body temperature, and other critical bodily functions (NIH website, https://www.nih.gov/).
- Restrict Important activities to high energy/hi focus times and Guard those times.

Maintenance Reality: The reality of the field of maintenance is that frequently we are not in control of our time. When we are not in control of most of our time, it is doubly essential to control what's left.

25 Meetings

Techniques to Make Meetings More Effective

Become an expert at both running meetings and participating in meetings. Expertise will make your life easier. Meetings are the bane of the supervisor's existence. Yet they are the best way to communicate with groups that do not work together or to convey new information. They are also the best way to collaborate to solve problems. They also can be used to fulfill legal requirements.

- Hold them standing up. Comfort increases the length of the meeting.
- Remember the rule (for the same number of agenda items): Coffee/soft drinks increase the length of the meeting; doughnuts increase the length, and food really (really) increases the duration of the session. But sometimes serving food or drinks is part of the meeting design.
- Write and distribute an agenda (consider have times on the agenda) and consider appointing a timekeeper.
- Require on-time attendance. Make a direct request of people who come late.
- End the meeting when you say you will, even if the agenda is not complete.
- Create a system to manage promises, requests, reports, inquiries, and research from the last meeting. These things must be available to refer to at subsequent meetings.
- Insist that everyone has done their homework from the last meeting. Do this by asking directly if it was done and get a promise if it was not done, that it be complete by the next meeting. Consider canceling a meeting if a quorum has not done the homework.
- If the meeting is called by someone else, try not to get invited if it doesn't concern you or your workgroup. The other strategy is to delegate attendance of the meeting to someone else on your team.
- Located on the first page of the minutes at the top if sent by email. Should show up in preview pane of email program.

Date activity assigned	The request or promise	Person(s) who has agreed	Date promised
3/7	Draft budget for discussion	Mary M.	3/15
2/28	Audit completion	John P.	3/11

Name of Organization or Group:
Purpose of Meeting:
Date/Time:
Chair:

ACTION MANAGEMENT		
Action	Person Responsible	By when
This item should be written in clear action language so there is no doubt what was volunteered, decided or assigned.	It is important to have someone who is accountable for each action item.	Next meeting? Date?
Next action - ordered by date or priority		
Next...		

Types of Meetings that You Might Find Yourself Running

TEAM MEETINGS

Team meetings should follow good meeting routines to maximize the value of the time spent.

- Have an agenda
- Agree to ground rules for the meeting)
- Take simplified minutes such as just action plans
- Set time limits (half an hour meeting or something)
- Have a chair who leads the meeting even if the position is rotated.

Agendas can be as simple as a couple of items as long as the meeting is at a set time and place. The reason for the meeting is to support everyone's Orbits, bounce ideas around, ask for help, brainstorm, and develop success stories.

Some possible team ground rules include:

Teams often develop ground rules for the meetings. Meeting ground rules could include agreements such as:

- Start and end on time.
- Attend all meetings. Be on time.
- Do homework.
- Value the diversity of team members.
- We will keep the focus on our projects and the issues surrounding their goals, avoiding sidetracking, personality conflicts, and hidden agendas.

- The integrity of your workgroup or team is similarly undermined when key people are missing; it doesn't matter why.
- Keep to the current topic, and we will avoid side-bar discussions, while others are talking.
- Thank people for attending and for their contributions to everyone's goals.

Agile techniques call for frequent, short, scrum meetings. Periodic check-in where each team member answers the questions. These meetings might include other topics, but they always include the following:

- What did I do since the last meeting?
- What I plan to do by the next meeting?
- What roadblocks are in the way?

TOOLBOX MEETINGS

Toolbox meeting is (usually) a quick morning meeting to discuss the work of the day. It is also designed to highlight some of the unique hazards of the day (such as crane lifts). Each meeting has a plan, attendance list, and is signed off by the provider (facilitator or employee leading the session). Records must be kept to qualify as a safety meeting and made available to an OSHA inspector on request.

Some of the topics for a toolbox meeting might include:

- Safety items of the day, such as lifts, hot work, etc.
- Some organizations have a safety moment to begin every meeting from topics below or timely topics (such as current events).
- Essential production items for the day
- Company announcements
- Personnel announcements and vacations
- Status of permits
- Break-in maintenance jobs
- Scheduled maintenance jobs
- Important interfaces between crafts
- Education minute
- CMMS reminders and training

SAFETY MEETINGS

In the US, holding periodic safety meetings is mandated by OSHA (the Occupational Safety and Health Administration). OSHA "requires that all companies hold regular safety meetings for both management and employees. The type of meeting that a company must hold to comply with OSHA regulations varies depending on its size and industry; however, companies must adhere to the safety meeting requirements to remain in business. A company found not in compliance with OSHA regulations may face a fine or have its operating license revoked." (www.ehow.com)

These meetings must be scheduled and held on company time. The topics would include at least the hazards that your organization faces. The meeting could even include home hazards and driving hazards. In factories and construction sites, the safety meeting can be tagged onto the toolbox meeting. This is (usually) a quick morning meeting to discuss the work of the day. It is also designed to highlight some of the particular hazards (such as crane lifts). Some of the topics for a toolbox meeting might include:

Safety meetings are an OSHA requirement. Pick your topics from the list below.

- Fatigue
- Fire
- Forklifts
- General
- Emergency procedures
- Safety tours
- Behavioral safety
- Confined space entry, working, rescue tools and safety watch
- Disease prevention
- Electrical hazards
- Ergonomics
- Hearing conservation
- Ladders
- Hazard communications: SDS (formally MSDS) sheets, understanding, labeling and the like, Training in the chemicals in the immediate workplace
- Topics about lifting and general back care
- LOTO: Lockout and Tag out associated toolbox topics
- Reporting incidents, near misses, hygiene
- PPE: Questions involving personal protective equipment
- Seasonal toolbox topics dealing with weather, annual issues, or holiday awareness
- Behavioral safety confined space disease prevention electrical hazards
- Ergonomics hearing conservation ladders
- Office issues, permitting
- Slips, trips, and fall
- Small tools topics associated with small hand or powered tools
- Weather and how it affects safety

Each meeting has an agenda, attendance list, and is signed by the provider (facilitator or employee leading the session). Records must be kept and made available to an OSHA inspector on request.

SUPERVISOR RUN PROJECT AND SHUTDOWN MEETINGS

Private Meeting Guide

If the main thrust of the project meeting is to be brought up to date (or have the management team brought up to date), you might prefer to circulate and talk to people informally, rather than hold a more formal meeting. Or you may prefer to circulate and communicate with selected people beforehand to get a sense of how they feel about the project, or what their current level of knowledge is. These conversations should take place before an update or status meeting so that any issues or discrepancies that are uncovered can be aired, discussed, and hopefully resolved at the meeting.

Discussion items for these informal talks may include:

- Do you foresee any problems coming down the pike?
- Is there anything else bothering you?
- Is there anything I should know privately about the project? These would not be discussed in the review meeting directly but are taken on background.
- Is your team working productively? Are resources being moved away to do other jobs? Will that impact your job?
- Is your team 'happy,' and do they need anything else to work productively? Is there anything missing that would make your job easier?
- Have you or your team had any ideas that might apply to other groups on this project?

A sample Project Review Meeting agenda (e.g., for a planned physical asset shutdown)

1. Begin with a review of the safety or product quality issues
2. Mention any lean or sustainability items, or environmental concerns which you may be trying to get people to think about for the long term
3. Scorecard: A brief review of where we are today to the project schedule
4. 5-minute or fewer reports from accountability holders–updates and reviews
5. Discuss accomplishments and acknowledge good work
6. New business (problems or opportunities not presented before)
7. Detailed view of breakdowns and mitigation efforts, including requests for resources or other items needing discussion
8. Review of action items and promises made at the last project review meeting
9. Quick overviews of major breakdowns brewing and hot upcoming topics
10. Plans for today
11. End with safety goals and issues

Essential: Maintain an action promised list, perhaps on a shared resource such as a web-based file accessible to everyone. This list reminds people of what they promised to do and increases the level of accountability. Meetings run more smoothly, and projects move forward with less acrimony wherever a workgroup maintains an action promised list.

26 Legal Issues for Supervisors

You perform actions that have legal consequences for your organization. Yet few people are trained in the law as applied to maintenance efforts. You might negotiate contracts, purchase materials, and services and even represent your organization in court. Some liabilities and some forms of negligence might flow to you personally.

CONTRACTS

For many maintenance departments, contracts and contractors are an essential component. It is vital to develop a basic understanding of contract law.

If your contract is based on delivering results, you can create a win-win situation for yourself and the contractor. In most shutdowns, results should be in the following order of priority

- Safety and environmental
- Reliability of equipment.
- Cost of delivering reliability.

If there is an incentive for a contractor to deliver reliability, it naturally follows an incentive to prevent a shutdown and to perform preventive shutdown, plan shutdown, schedule shutdown, and so forth. In summary, they need a disciplined process in place and a good system to support it.

When Is a Contract Needed?

(Caveat: Discussion that follows is based on US commercial law. Most countries have similar provisions. Be sure to verify local law with company lawyers before proceeding. Partially adapted from the work of Mike V. Brown, President of New Standard Institute).

A contract should be drafted whenever construction or other services are required. The form of the contract can be informal, such as a purchase order referencing a proposal by the contractor. It can also be very formal, requiring bids along with detailed descriptions of the work and expectations of how the work will progress. In some states, all work on real estate must have a written contract.

Note: For purposes of contracts, a corporation is considered a person.

First, what is a contract?

- A contract is a voluntary and deliberate agreement between two or more persons constituting an offer to receive and pay.
- Unconditional acceptance required.
- Can be verbal or written but in some localities must be written if over $500.
- Can be conveyed by purchase order or sales agreement.
- Acceptance in writing or by performance.

RULES OF ALL CONTRACTS

- Contracts can only be entered by sane adults.
- No contract can violate the law.
- For a contract to exist, an offer must be made:
    - Not a sales brochure or catalog listing.
    - Must be specific to actual goods or services.
    - The purchase order form is an offer vehicle.
    - The sales agreement form is an offer vehicle.

  and acceptance
    - Acceptance must be unconditional, or no offer exists
    - Example: Supplier offers to sell a gearbox at $1000, but the Buyer offers to pay only $900. *Offer No Longer Exists*. If Supplier counters with $950, and Buyer says OK at $950 then (and only then), *Acceptance exists*
- UCC (Uniform Commercial Code in the US) allows acceptance with some changes:
    - "as is."
    - "buyer objects to any added terms."
    - "payment does not end buyer's rights."
    - "The price is assumed to be firm."
- There must be an exchange of goods and or money.

TYPES OF CONTRACTS

Contracts for services take several forms. Lump-sum and unit-price contracts are commonly used because of bids submitted by more than one contractor. A company can let negotiated contracts to one or more firms. The contractor may be chosen, not so much for the price, but rather for dependability, experience, or skill.

AGENCY

An agent is a person who is able to make decisions and take actions for, or that impact, another person or entity like their employer. Companies appoint agents to carry out essential processes with outside firms. Ever wonder why a purchasing person is called a Purchasing Agent? They are legal agents of the organization and can bind the company with various vendors into contracts.

Agents are appointed "officially" like the purchasing agent. There is a contract up that outlines the powers, grants of authority (how much they can spend), and other rules.

But agents can also be appointed by circumstances, even informally. One set of conditions is the past practices of people in that role. If they made purchases or hired contractors, then a vendor might assume that you can too. In agency law, if you act like an agent, your organization pays bills for what you contract or purchase, then under the law, you are an agent.

As a supervisor, you may become an agent of your organization. For example, if your organization allows you to buy and pay for maintenance spares, an outsider can assume that you are an agent. Once you are an informal agent, your word would bind the organization.

The problem comes in when the agency is informal. According to the Law Dictionary, "The extent of each parties' liability for an act is based on whether an agency relationship was created and the scope of the agent's authority to act."

As an employee, you should discuss with the HR department or Manager as to the limits of the agency relationship.

27 Effective Decision-Making and Delegation

Making the right decisions with inadequate information, resources, or time is the sign of mastery of the supervisor's environment. You will make mistakes. Hopefully, you will learn from them. After you make a decision, think through your process and see where you might have improved. Given the same information, would you make the same decision again?

- "The success or failure of most businesses is determined by the sum of hundreds of small, good decisions made by lower management people and the relatively few major decisions made by top management."

  –Gail Crawford, Executive VP Ringsby Systems

The importance of this section was stated very clearly in an article published over 45 years ago called "How Profitable Decisions Are Made," by Gail Crawford, Executive Vice President of Ringsby System:

"The success or failure of most businesses is determined by the sum of hundreds of small, good decisions made by lower management people and the relatively few major decisions made by top management."

Maintenance supervisors make hundreds of decisions that will have direct, indirect, short-term, and long-term effects on the organization. This section helps you in your decision-making and (possibly more important) clears the decks of decisions that should be made by others to give you time for the more critical decisions.

Most supervisors have limited time. Decisions have to be made on a rapid basis, frequently without adequate preparation. This section describes several methods to improve your decision-making efficiency.

Frequently, making the decision IS the issue. Many decisions must be made, and the rightness/wrongness is not at issue. This is especially hard for newer supervisors because your crew's total productivity is frequently tied to your ability to decide the course of action.

Walt Lacy, a leading trainer in this area, makes the analogy that making decisions is like trying to steer a car without power steering: It's tough to steer until the vehicle is moving. Once the vehicle is moving, it's easy to steer. Once any decision is made, it's much easier to change course.

When decision-making is a problem, how does the problem show up? Review your personal style inventory and relate your style to the three avoidance areas below:

- Disorganized: Too many decisions to make; don't have time.
- Perfectionism: Constant worry about decisions, you never know when you might make a mistake.
- Procrastination: Put off making decisions.

GUIDELINES FOR DECISIONS

Establish the goals of your workgroup. What is your group's mission, the reason for existence

Guidelines for better quick decision-making:

- Your choice should be in alignment with the mission, vision, and values of your organization.
- Your choice should alter the work group's goals the least.
- Set aside your typical pattern for delegation.
- Take control.
- Take full responsibility.
- Act.

This is not the time to observe all the niceties of discussion or consensus-type supervision. You must take control, take responsibility, and act. Maintenance emergencies frequently require this type of decision making.

TEN STEPS TO BETTER LONG-TERM DECISIONS

1. Define the problem. Is there a decision to be made?
2. Collect knowledge. Knowledge should pass the "BAR." (Brief, Accurate, Relevant)
3. List all possible alternatives; involve as many members of your team as makes sense.
4. Prioritize, choose the best one, and make sure the choice is in line with your overall goals.
5. Don't discard the other ideas; they can be used as contingency plans.
6. Inform all members of your team what the change means, why it was chosen, how it will work.
7. If a change to a formal procedure is required, tell everyone what to do and give it time to sink in. Hopefully, members of your team were part of the decision process.
8. Action.

9. Monitor progress and the achievement of milestones.
10. Compare outcomes with goals for your workgroup. Did you achieve your goals and increase quality, efficiency, safety?

DELEGATION

Working through other people is the core activity of supervision. Your ability to inspire, to get cooperation, to get work done, is the ultimate measure of your success.

"Supervisors are not paid for what they can do but for what they can control." (stated by Lee Minor, teaching "How to Supervise People")

According to Webster's New World College Dictionary, delegation is defined "to entrust (authority, power) to a person acting as one's agent or representative." The first ground rule of delegation is to entrust both the authority (power) and the responsibility.

The supervisor's job is to work with other people. One of the most challenging transitions is from the worker (being paid for how well you work) to the supervisor (being paid for how well you work through others). One of the supervisor's primary jobs is the development of the people that work for them. The delegation will help develop talent within the workgroup. The effect of responsibility on people can be amazing.

Resistance: Many supervisors resist delegating work to their subordinates. Often the reason is that the supervisor is sure they can't handle it or he/she can do it better.

Other fears might be:

- Bad quality might reflect poorly on the supervisor
- Loss of control
- Feel threatened by training a replacement
- Look bad because other people are doing the "actual" work

Supervisors often like their jobs because they want to feel needed (always being the center of attention). Delegation might seem like it reduces their position.

On the other side of the coin, the subordinate might resist the assignment because they feel as if they are set-up, they are already too busy, or they learned it's safer to rely on the supervisor.

Do's and Don'ts of What to Delegate

DO's	DON'Ts
1. Routine tasks	1. Personnel tasks
2. Time-consuming jobs	2. Job assignments
3. Skill improvement tasks	3. Disciplinary actions

Delegation is an opportunity for coaching. Allow enough emotional space for your people to grow. Keep yourself from interfering (we know you can do it better, but they need to learn how to do it as well as you). If they are not in danger (to themselves, to you, large batches of product, etc.), let them learn, it will make them better employees. Consider a replacement supervisor as a training ground for future supervisors, which you will need when you vacation (or go to seminars).

The One Minute Manager, by K. Blanchard, PhD, and S. Johnson, MD.

The One Minute Manager, which came out in 1982 and still going strong, is one of the most popular self-improvement books of all times for managers. The rules apply directly to maintenance supervisors. We strongly encourage maintenance managers to purchase and read this excellent book (new version in 2016). Three concepts are simple and powerful. The issues are productivity, job satisfaction, and the effectiveness of the workgroup.

To maximize the effect of *The One Minute Manager* concepts, discuss the goals and ideas with your workgroup first. Reprint this page for everyone to read and encourage people to read the book (supply some copies). Discuss the concepts, so everyone knows what the score is.

1. **One-minute goal setting:** Make it clear what the subordinate is to do. Write the goals out on a single sheet of paper. The complete statement should be less than 250 words. In the words of the author, "feedback is the breakfast of champions." The goals should be written in the first person (using I), and in the present tense. ("I arrive at work on time every day." "I keep my safety equipment in good repair.") They should read their goals every morning (it should take less than one minute).
2. **One-minute praising:** Catch people doing something right! It doesn't have to be perfect; look for anything that is approximately right. Remember, feedback is the breakfast of ______.

   Steps:
   1. Tell the person what they did right.
   2. Tell the person how that makes you feel (e.g., happy, proud).
   3. Do it now.
3. **One-minute reprimand:** This is an opportunity to express your anger and frustration before it can build up and become destructive. Your employee may have reasons for what they do, but you won't find out what they are if you blow up at them. They may even appreciate hearing from you about what is expected of them. Idea: We are not our behavior. Before you reprimand someone, be sure you have the facts.

Steps:

- Give the reprimand in a private location.
- Be specific, tell people exactly what their behavior was that made you angry or uncomfortable.

- Tell them how that behavior made you feel and why.
- Allow a pause, make sure the person understands that their behavior was the issue, not them as a person.
- Do it now; don't wait.

Three thoughts on *The One Minute Manager*

1. Beware: Do you believe that you must be great at it for it to work? The fact is that it will work if you use it. You will improve over time.
2. This new tactic will increase productivity. The better your people look, the better you look.
3. Good productivity is a journey, not a destination.

28 Trade Training

The gap that is critical for maintenance is between your team's skillsets and the skill requirements of your machinery, tools, and other assets. The skills needed to run today's factories and buildings are changing faster than some people can assimilate. Jumps in technology disorient even the most dedicated worker. One of your jobs is to realize there is a gap and manage the processes to fill the gap.

CASE

In a factory manufacturing fuel control system—Chief of Field Service, Calvin Smith, was 55 years old. His workforce consisted of three younger technicians. He was required to do service himself and usually took (or was brought) the most challenging problems. He was the best and most highly skilled troubleshooter for over 12 years through the transistor era (he actually started with relay logic).

The company moved to CMOS integrated circuits. Now each integrated chip replaced an entire board. After a painful learning process, he came up to speed on CMOS. He never developed the comfort level with integrated circuits that he had had with transistors. Instead of knowing and following the entire logic of the board, he associated specific chips with specific faults and replaced the chips until the board worked. But after two years, his expertise was quite good. His attitude had recovered.

Then the new microprocessors started to show up in the designs. He started over, but the difference between microprocessors and CMOS was much more extensive than the difference between CMOS and transistors. He never understood the concept of programming or the dynamic nature of the data and address bus. He couldn't understand how to troubleshoot a dynamic system, such as a typical microprocessor board.

Our dedicated field service manager took very early retirement, feeling that the world had passed him by. The company lost his expertise. He lost his sense of mastery and feeling that he was part of an important field. He now does odd jobs (he started as an electrician) in his neighborhood. Interestingly, his old subordinates still bring him certain problems which he throws himself at with relish.

This was a waste of human resources. Proper training would have saved this person and his highly valuable expertise. While the company spent money to train the engineering staff, it didn't think to include the service staff. The design lag time (it took two years to get the first microprocessor product off the drawing board) would have been more than adequate to train the entire staff.

Always ask yourself the question; does this person need training or counseling? Many attitude problems stem from inadequate training and the resulting feelings of incompetence. Thus, many attitude problems can be resolved quickly, even cheaply.

Training That Hits the Target

How do you design training the fills a gap in someone's competency?
How do you conduct training that is not boring?
How do you ensure that the training doesn't fall on deaf ears?

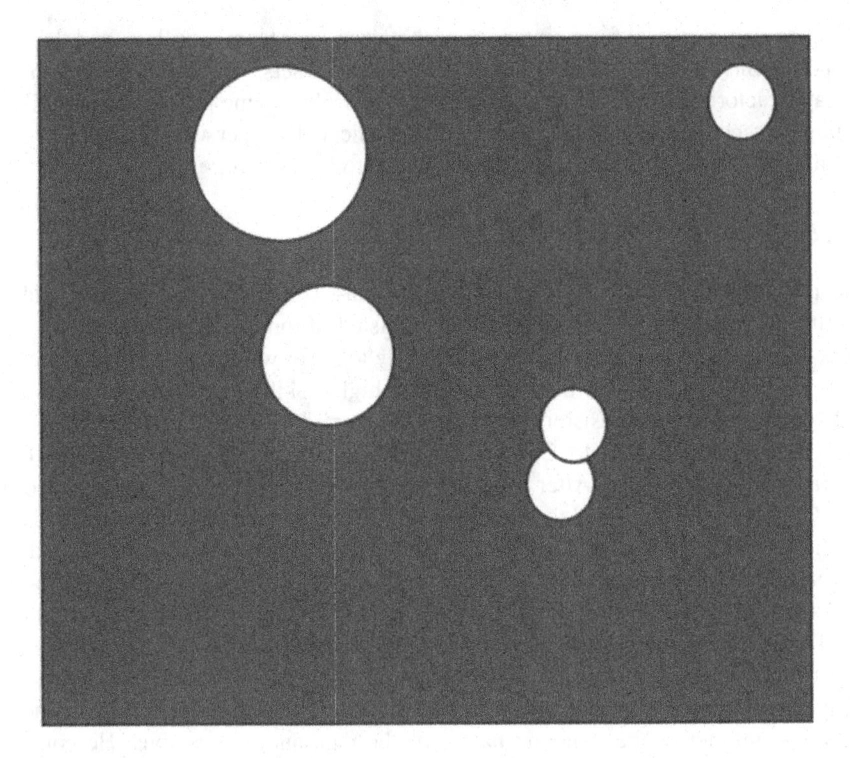

FIGURE 28.1 The mind before training.

This is an essential set of questions. I would like to present a model for competence in any knowledge or skill area. The square represents the field of all the knowledge or skill needed to do the job (whether it is welding or a reliability engineering position).

The white circles are holes in the person's competency. If we started with a young person just out of college or trade school, we would find many holes. If we would revisit the same person five years later, many of the holes have been filled in with experience, ongoing schooling, and iOJT (intentional On the Job Training).

If the environment is a rich one where the new employee can get a wide range of experience, then they will have more holes filled in by experience. Conversely, if the environment is not rich, then a few holes would be filled. That is the reason that apprenticeships are so important. They insure for the first few years, at least, the person is in an intentionally rich environment with alternating working and classroom assignments.

An old-timer will have most of the circles filled for their present environment if that environment stays stable. Once the situation is in flux, then the old timer's knowledge and competence base gradually become less aligned with the field of competency needed for the job. Their competence (that was so personally hard and time-consuming to obtain) becomes obsolete.

The other interesting thing is people who have worked in the same facility for their career have their competence field filled in for that facility. The competence issue only presents itself when they change jobs. They might be entirely competent with the equipment, tools, and processes of their old job but might (will) have holes when presented with the new situation. Some of these holes can be dangerous because they are usually a surprise to the person themselves. Since the person was competent in his/her old position, they develop a habitual way of thinking about themselves as competent and don't question it.

Where and How the Training Is Positioned in the Field Is Essential

In the simplified diagram, we can see three different trainings (A, B, and C). It is important to realize that this discussion is independent of the quality of the trainer or the training materials, the modality, even the willingness to learn (they can significantly help mitigate this problem but don't change it).

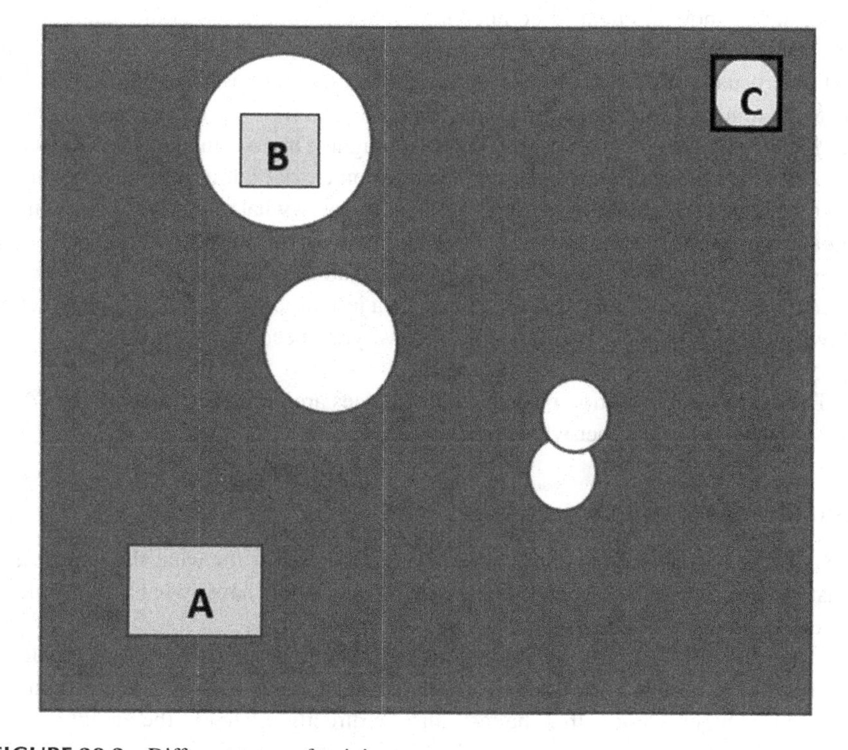

FIGURE 28.2 Different types of training.

Training A is a review of the knowledge and skills that the candidate already knows. There is a high probability of boredom since it cuts no new ground. Of course, an excellent instructor and engaging materials will help, but the tendency toward boredom is there.

Training B is material that is not connected to anything the candidate already knows. It is too advanced for this candidate without some preparatory work. Someone in this position is bored, frustrated, maybe annoyed (even at themselves for not knowing it or at the trainer, materials, or company for not being clear). Like walking into an advanced class, you will get very little out of the experience. As before, a gifted instructor might be able to backfill enough material, so the main topics make sense.

Of course, **Training C** is ideal. Some of the material is known to the candidate. Also note, all the new content is linked to the areas of competence. Elementary school teachers, in their wisdom, repeat known materials as they transition into new topics. New topics are always tied to known topics, and where the material is entirely new, the teacher proceeds slowly and repeats often. Of course, with a large classroom, there will be boredom since some students will "get it" the first time and frustration because some students won't even "get it" the 10th time.

Where Does This Leave Us?

In the maintenance world, we must get over our fear and dislike of testing. Testing is important to show the field of competency. Without knowing what the candidate is competent in, training becomes a hit or miss proposition.

Firms usually train everyone on some course they think is essential. This is like throwing mud at a wall and seeing what sticks. It does have the advantage of appearing "fair" and of not needing a lot of forethought. One advantage of experienced trainers is that they are used to this and can, to some extent, mitigate the problem.

If we have scarce training dollars, we need to know what competencies are missing and aggressively go after them. Training this way is not easy and requires more effort from the staff for testing, training design, and thinking. It also will not be fair since the goal is a full field of competence for a job, process, or area, which depends on where one has started. In this model, not everyone needs the same training, so not everyone gets the same training.

The unanswered question is, what competencies are needed for a particular situation? That could be another whole discussion.

You Must Deal with Aptitude

A new supervisor needs to evaluate all their subordinates for what they know and what they do not know, what skills they have and don't have, what attitudes they have, and finally, what their aptitudes are.

Underlying everything is individual aptitude. Some of the performance problems observed by supervisors are aptitude problems. Testing can usually uncover aptitude problems. The problem is that the test must be directly related to the activity of the job itself.

According to Wikipedia, an aptitude is "an innate, acquired, learned, or developed component of a competency to do a certain kind of work at a certain level. Aptitudes may be physical or mental. The innate nature of aptitude contrasts with achievement, which represents knowledge or ability that is gained."

What aptitude, when missing, could cause slow production or quality problems?

- Lack of strength (in the trunk, hands, arms, legs)
- Lack of endurance
- Lack of flexibility or agility
- Lack of visual acuity
- Cannot hear instructions
- Lack of adequate intelligence

Aptitudes are innate, so training will not make up a large gap between the needed aptitude and the candidate's underlying ability. If the lack of aptitude is slight, then if someone wasn't strong enough, they could train; if they had visual acuity deficiencies, they could get their vision corrected.

The other choice is accommodation. If strength is the issue, perhaps some handling equipment can be used. Maybe the controls can be moved if the candidate is not flexible enough to reach them and the part they are working on. Perhaps another team member can swap specific responsibilities with a candidate who cannot perform that task.

It is the responsibility of the supervisor, workgroup leader, foreman, or gang leader to notice aptitudes and take the appropriate actions.

A useful and straightforward guide to the different levels of mastery of a skill or knowledge area:

Phase 1: Do not know.
Phase 2: Know the theory but cannot do it.
Phase 3: Can do but cannot teach.
Phase 4: Can do and teach.

Competency

Corporate trainers think in terms of three types of learning: Knowledge, Skills, and Attitudes. For supervisory jobs, a person in charge should be competent in all three areas to be effective. Many types of training address one or two of these types of learning without regard for the other. For maximum effectiveness, all three areas require addressing.

Types of Competency	Observable Behavior	Performance
Knowledge	Be able to describe, diagram, or argue.	Answer x of 10 questions correctly.
Skills	Be able to demonstrate, show, perform, solve.	Do … in x minutes with no mistakes.
Attitude	Act with comfort, without hesitation, without grievance.	To your satisfaction.

Let's analyze the previous example from a competency perspective.
Calvin needs competencies in all three areas to be an excellent chief technician. He lacks specific knowledge that could show him what is happening on the board. He requires some particular skills to help him fix the boards. His biggest problem now is a negative mental attitude. His attitude stands in the way of his gathering the skills and knowledge. The attitude will be the last to be fixed and will follow mastery of the skills and knowledge.

There may be a fear here too. He may fear that he no longer has the aptitude to learn this new material. This is a defensive fear (a fear that protects the person from finding out unpleasant information about themselves), which is common, especially when the old knowledge was hard-won.

STEPS TO DEVELOP TRAINING AND TEST THE NEED FOR RETRAINING

There are four steps in the design of a tailored learning program:

Step 1: Determine what knowledge, skills, and attitudes are needed for the job. Before we can investigate teaching anything, we must see what competence is required for the job in question. Look at the situation as it is today and forecast where the job is going soon. The big picture is the General Learning Objectives (GLOs), deciding what needs to be accomplished. The concrete and specific skills, knowledge, and attitudes required to do the job are called specific learning objectives or SLOs. In a correctly designed learning program, people achieving these SLOs will be successful in their jobs.

What is needed to troubleshoot microprocessor boards? The organization must decide what level of competence is appropriate for a service chief in servicing these boards. A form should be filled out and kept with job descriptions for the Chief's job for future reference. Note that this is the microprocessor part of the job only. The chief's job has many other facets.

Step 2: Evaluate the potential trainee's (or trainee groups) current skills, knowledge, and attitudes. A direct supervisor might be able to make an educated guess. If the trainee has good insight, they might know where they are weak. Most situations require testing (either observation on the job, or more formal written or bench tests). The examination should be designed to uncover the competencies on your required list from step 1.

It is important to note that success on the test should correspond to success on the job. Testing that does not reflect job requirements is said to be invalid. In the USA, the Americans with Disabilities Act (ADA) and related legislation are clear that the test must not discriminate against any group, disability, or condition. For example, if the worker must lift 100 pounds in the trial, the job must call for heavy lifts where equipment cannot easily be used. In the US, "reasonable accommodation" must be made for people with disabilities.

We must evaluate our service manager's skills, knowledge, and attitudes. After we make a list of specifics, we rate (or test) him/her in each area. The result, tailored to this individual, would be kept in the personnel or training file for that individual.

Step 3: Translate the gaps in skills, knowledge, and attitudes of the potential trainee from the required list to develop a training lesson plan. The training plan should list all types of learning that this person/group needs. Summarizes the skills, attitudes, and knowledge that Calvin Smith lacks and needs for the job. The form should recommend possible exercises and resources to provide the learning that Calvin needs. The form also estimates the time required for the trainee and the need for any supporting staff.

You can go through this exercise for all related jobs. Other service technicians might have related SLOs; it is more economical to train people together wherever training goals are compatible.

Step 4: Posttest or retest. Test to determine if the training effort was successful. If not, put the candidate back through the process or determine if they don't have the aptitude for the job.

Training Function in a Maintenance Department

Training is an essential issue if a maintenance department wants to maintain high-quality standards and decent morale. Any organization that doesn't invest in ongoing training is making a costly mistake.

In some cases, training must be a grassroots issue handled within the Maintenance Department. Supervisors might take it upon themselves to begin an ongoing training program by bootlegging resources from other areas, jobs, vendors, budgets, or interested parties.

The training function involves the following:

- Set policies, priorities, and goals.
- Check present systems of education and training.
- Look at existing certificate programs.
- Establish training systems for maintenance skills.
- Prepare a training calendar.
- Prescribe training based on task analysis and existing candidate competencies.
- Train the employees, upgrading the skills, knowledge, and attitudes needed, as determined by examination.
- Post-test candidates to be sure materials were learned. If not, adjust the approach.
- Evaluate activities and adjust the plan.

After the data is collected, you must determine where each of your workers needs help. Once that is done, build a to-do training file for each person.

Quick questions for your current training practices

Is there a present system of education and training for operators?

Do you have a written SOP with steps for educating and training activities.

Are there set policies, priorities, budgets, and goals?

Are employees trained for upgrading the TPM skills determined by analysis?

Is training based on task analysis and existing candidate competencies?

(*Continued*)

Are candidates routinely post-tested to be sure materials were learned? If not, adjust the approach

Review existing certificated operator programs or operator levels that exist.

Is there an ongoing evaluation of activities and study future approaches?

Actions to take

Prepare training calendar if one is not present.

Establish a training system for operation and maintenance skills if one is not present or not adequate.

Think about asking someone without a lifetime of maintenance experience to do basic maintenance tasks and solve problems. Some of the background knowledge is not there and cannot be assumed.

Education versus Training

There is a difference between education and training. Training is generally associated with specific skills and techniques. You can be trained to weld or how to wire instrument panels. Education concerns the whole person.

Education, in its broadest sense, is any act or experience that has a formative effect on the mind, character, or physical ability of an individual. In its technical sense education is the process by which society deliberately transmits its accumulated knowledge, skills, and values from one generation to another through institutions. (Both definitions from Wikipedia.)

Teachers in such institutions direct the education of students and might draw on many subjects, including reading, writing, mathematics, science, and history. This process is sometimes called schooling when referring to the education of youth.

The term training refers to the acquisition of knowledge, skills, and competencies as a result of the teaching of vocational or practical skills and knowledge that relate to specific useful competencies. It forms the core of apprenticeships.

One can generally categorize such training as on-the-job or off-the-job:

On-the-job training takes place in a normal working situation, using the actual tools, equipment, documents, or materials that trainees will use when fully trained.

Off-the-job training takes place away from typical work situations. Off-the-job training has the advantage that it allows people to get away from work and concentrate more thoroughly on the training itself.

You go to a university to get an education, and you go to a trade school to get training. Of course, the two categories overlap. Topics such as reliability are better dealt with within an educational framework and others such as tightening bolts in a training framework. They also have implications for the act of training. Training is done either on the job or off the job, while education is generally off the job.

In the lists below, many topics are partially shared. For example, safety is an area of education, while safe work practices are considered training. The former is theories, attitudes, the background to help understand the whys, and the former is concerned with specific methods.

Education-type topics (more knowledge-oriented):

- RCA (Root Cause Analysis)
- Reliability

- Criticality Analysis
- Quality
- Safety
- P/F curves
- Asset Management
- Replacement economics
- Life cycles of equipment and mechanisms of breakdown
- How your production works, how your industry works
- Preventive Maintenance
- Maintenance Management
- Why use different maintenance strategies?

Training-type topics (skills)

- Tightening bolts
- Lubrication
- Cleaning
- Inspection
- Measuring
- Sketching
- CMMS usage and filling out corrective work orders
- Specific adjustments
- Safe work practices
- Tool use
- Inspection of product
- Minor repairs

Organizing the Training Effort

Your department (if not the whole company) should set training goals for all craftspeople. Look at where they are now, where you need to go, and what is missing.

Budget at least 1% (20 hours per year) for training each person. Firms in rapid change might need as much as 5% (100 hours/year) or more to keep competencies and morale high.

The training director can even be one of the people in the department; the function could rotate among different people every year. Set-up files for each competency you envision needing training. For example, if you have Allen Bradley PLC's, how are you going to keep people up to speed? What are the elements for success with that technology?

Set up files for each person with the GLOs that they need. Review the GLOs with available time, business cycle, and funds. Be sure to act with the full input of the worker and supervisor. Once underway, review each person's file every six months to a year and be sure everyone gets an opportunity to be trained.

In addition to the craft-specific topics for training, some non-craft issues which should be considered are quality, safety, CPR, firefighting, toxic material handling, toxic waste regulations, your maintenance information system, statistics, filling out paperwork, PM, scheduling, project management, report writing, shop math, drafting, CAD, computers, engineering, cost accounting, your industry, your end products, what it's like to operate your machines. The list is endless. Your people will be better for the attention and the training

22 Guideposts for Adult Training

There are many "rules of thumb" in the teaching of adults. The more rules the supervisor follows, the more likely the training will be successful. Larry Davis' leading book on adult training and education, *Planning, Conducting, and Evaluating Workshops,* * gives 22 rules for teaching adults which sum up the best thinking on the topic:

1. Adults are people who have a good deal of firsthand experience. Effective training taps into the adult's existing store of experience.
2. Adults are people with relatively large bodies subject to the stress of gravity. Practical training allows adults to take breaks, move around, and change pace.
3. Adults are people who have set habits and strong tastes. Effective training is sensitive to adult habits and tastes and tries to accommodate as many as possible.
4. Adults have some degree of pride. Successful training is careful with the egos of the participants and helps develop greater abilities and independence in the areas being trained.
5. Adults are people with things to lose. Proper training is concerned with gain and not with proving inadequacy. The most effective training has 100% success ratios.
6. Adults are people who have developed a reflex toward authority. Good trainers (and good supervisors too) know that each adult has a different style of dealing with authority and doesn't take any of it personally.
7. Adults are people who have decisions to make and problems to solve. Effective training is problem-solution oriented and entertaining.
8. Adults are people who have a great many preoccupations outside of a particular learning situation. Effective training does not hog the adult's time. Training should achieve a balance between tight presentation and time needed for learning integration.
9. Modern adults may be bewildered by all their options and opportunities. Effective workshops assist them in selecting what is currently essential.
10. Adults are people who have developed group behaviors consistent with their needs. Effective training concerns itself with satisfying these needs and allowing many different responses.
11. Adults are people who have established emotional frameworks consisting of values, attitudes, and tendencies. Training denotes change. The change puts people's framework at risk. Productive training assists adults in making behavior changes. Active training assists adults in becoming more competent.
12. Adults are people who have developed selective stimuli filters. Effective training is designed to penetrate the filters.

13. Adults are people who respond to reinforcement. Some respond to positive reinforcement. All occasionally need negative reinforcement. Effective training is built on appropriate reinforcement.
14. Adults are supposed to appear in control and who therefore display a restricted emotional response. Sometimes intense training loosens up these restricted responses. Effective training is prepared for emotional release if it occurs.
15. Adults are people who need a vacation or time off from work. Effective training provides some time away from the grind.
16. Adults are people who have strong feelings about learning situations. Effective training is filled with successes.
17. Adults are people who secretly fear falling behind and being replaced. Effective training allows people to keep pace with the field and grow with confidence
18. Adults are people who can skip certain basics. Effective training starts with where the adult is today and builds on that.
19. Adults are people who more than once find the foundations of their world stripped away. Effective training reminds them of their ability to learn and start again.
20. Adults are people who can change.
21. Adults are people who have a past. Effective training is concerned with the development of new competencies. The why's of the past are someone else's concern.
22. Adults are people who have ideas to contribute. Effective training leaves room for their contribution.

* This excellent text on adult teaching is available from University Associates, Inc. 8517 Production Ave., San Diego, CA 92121.

Sources of Training

Training is a big business for many organizations. Here are some ideas:

1. Your staff is your first choice of potential trainers. Within your team, there are several possible opportunities for trainers. Please note that being a trainer should be viewed as a job enhancing project. Time should be given for preparing materials. The trainer should be relieved of other duties.
   - Tap your soon-to-retire people as trainers. This group has significant experience that should be channeled into the next generation. In some organizations, people who have already retired are recruited to return as part-time teachers.
   - Use the internal guest instructor concept. With this concept, a staff member would be treated as a guest (receiving lunch, clerical support, go off-site for more extended training, get relief from other duties).
   - Tour training is an excellent team-building exercise. Once a month, you tour a section of your facility, and the most experienced maintenance person plays "show and tell" about the problems and successes in his/her area.

- Video technology has rocketed ahead so that a quality video camera is not needed, and editing software is commonly available. New equipment set-up, construction documentation, and specific machine training are popular first subjects. Craft training is a more difficult but rewarding area. After expertise is obtained, any topic can be video recorded to good effect; these can then be used to train new people. Shorter videos are more effective than longer videos. Many organizations create a channel on YouTube so that people can watch the videos at home.
- Look to other parts of the organization such as human resources, data processing, engineering, and production for expertise useful to your upgrading effort.

2. Many excellent companies provide craft training. These firms can provide professional instructors, testing rigs, and video/audio files or streams. The quality and appropriateness of your operation may vary, so check several vendors.
   - In-house courses are available on a wide variety of maintenance topics. Most appropriate if many people need the same training. Costs are about $1500–$4500 per day.
   - Public seminars are useful for training one to three people. Expect seminars to cost $500–$3000 and last one to five days. Try to get recommendations for better workshops from people in your industry.
   - Many organizations still sell DVDs. Expect to spend $50–$7500 for video training in a wide variety of areas. These same organizations also maintain a streaming library.
   - Interactive computer or web-based streaming videos can lead a trainee through a series of lessons and retrain when necessary. Expect to pay $50–$500 for a rental and up to purchase a course. These typically require only a standard PC, tablet, or other streaming devices.
3. Internet training. Some computer networks provide internet-access training sold by the hour. These systems train in electricity/electronics, pneumatics, building trades, business subjects, computer subjects, basic science, and many other areas. Try: www.onestoptraining.com
4. Equipment manufacturers. This is a growing area. Equipment manufacturers have a vested interest in a trained user base. Many of them subsidize training by calling it a marketing expense. Time and time again, it has been shown that trained users are happier users. Try to negotiate training into all equipment purchase contracts. Excellent low-cost training is usually available from vendors of predictive maintenance hardware.
5. Trade and professional associations. These groups strive to increase their value to their membership. One of the traditional ways is to provide industry-specific training in either traveling seminars or at workshops during trade shows. If your association does not provide training that you believe is needed in your industry, why don't you volunteer to put a seminar package together for the association?
6. Tech schools are an excellent source for trade training. Get to know the people running your local tech schools. Visit and walk through the facility. Many

companies set up specific labs, benches, or workshops with equipment that the tech school uses to train students. Many tech schools are willing to negotiate training contracts for some or all your technical training needs.

7. Community colleges, colleges, universities are frequently looking for new markets. These institutions have significant expertise in teaching more advanced subjects to adults. Many of them have entered instruction contracts with private industry in areas including computerization, robotics, regulation, automation, business skills, and other topics.
8. Some unions are rethinking their traditional roles. Many see that skill needs are shifting and have decided to lead the trend by setting up training for their members. This might be an interesting subject to raise if your union is not doing this already.
9. Insurance companies can cut claims by conducting certain types of training. Some firms will send risk managers through your facility and provide specific training in areas such as safety, risk management, liability reduction, fire safety, storage and handling of chemicals, record keeping for maintenance, safety, and accidents.
10. Government agencies provide seminars and workshops on a wide variety of topics, including EPA issues, hazardous materials, waste disposal, safety, record keeping, dealing with overseas vendors, and many others.

Methods to Consider for Training

- VR, Virtual Reality, a hot area that is coming into its own. The VR simulates an immersive reality where the trainee can manipulate reality and learn. Long popular with gamers and widely used by the military and in pilot training.
- AR - Augmented Reality where simulation overlays reality giving the wearer critical information.
- Coaching, iOJT (intentional On the Job Training): iOJT is one on one training and encouragement (especially useful if you use the technique TSED—tell, show, have them explain, have them do). This technique is suitable for teaching critical new skills where the high time investment is justifiable.
- Case method: Analyze a specific incident, problem, situation, or company.
- Books and reading materials: This media is the lowest cost method, suitable for people already skilled to add a specific skill or knowledge area. A permanent resource can be referred to in the future and used by others.
- Video and audio: These have the advantages of lectures, books, and demonstrations combined. A permanent resource can be referred to in the future and used by others.
- Conference: Send someone to a public conference with a training program. The trainee can get exposure to many instructors, peers, and vendors at the same time. And maybe have some fun, too.
- Demonstration: Trainer shows trainees how to do something, clarifying or highlighting the best way to do something.
- Laboratory: Experiments are designed to teach by discovery.

- Lecture: Trainer directly instructs trainees with the material to be learned. The lecture provides necessary information on a topic. It can introduce topics to many people at the same time.
- Programmed learning: Trainees go through the material at their speed. It can be accomplished through books or online lessons—accommodation for trainees who need additional content in some texts.
- Roleplay: Trainee plays a role in a simulation of real situations, and learns by doing as well as through the reactions of the other role players.
- Simulation: Trainee is presented with a realistic scenario, and the trainee works alone through problems and situations.
- Webinar: Using a computer and internet hookup to conduct training from a remote location. Several plants or buildings (even globally) can share a great instructor at the same time.
- Online degree programs: Are programs combined several different channels for learning and are managed from a portal (called an LMS - Learning Management System) for the program and class.

Checklist

If you run the training yourself, consider the following:

- Materials called educational software, which includes books, worksheets, etc.
- Staff time scheduling, including contingency plans, if key players cannot attend. Replacement on the shop floor. Consider people needed both inside and outside the training room before and after the training.
- You are contracting with outside firms needed for guest speakers, professional trainers, turn-key training, slides, video production, audiotaping.
- Trainees were invited and have sent in their RSVP. Be sure to avoid shutdown days where everyone is needed or the days after a shutdown where everyone is catching up on sleep.
- Have a plan for responding to various levels of emergencies so that everyone doesn't get disturbed for every little breakdown.
- The structure of training might include break-out sessions, hands-on bench work, access to computers, and classroom.
- Providing aids, including PowerPoint, flip charts, microphones, slides, videos, projectors, screens, satellite link-ups. For the best learning conditions, including comfortable chairs and tables for writing on, perhaps even access to power outlets to plug in laptops for taking notes.
- Facilities should be comfortable and large enough for the number of trainees expected. Who is the contact person? Who has the keys? How do you turn everything on and off?
- Having a contact person managing accommodations for trainees who are traveling in from another facility.
- Food and refreshments help make people more receptive. Remember, people have different tastes. Only 40% of people drink coffee in the morning; others may drink tea, water, juice—plan for your group.

- Check dates for conflicts with vacations, holidays, local holidays, hunting seasons, work schedules, bad weather.
- If people are traveling by air or train, coordinate pick-ups, tickets, vans, etc.
- Use promotional techniques to sell the program, persuading people to want to attend.
- Timing can make or break a program. A giant reorganization before training can be a good thing, or may ensure no one will have their minds on the material; a restructuring afterward can be devastating.

For reference: If you develop PowerPoint slides for your training, consider the following tips:

- With any training, know your audience, their level of expertise, and their expectations.
- Know the time of day and condition of the audience (after lunch, after work)
- Pace your talk. Overheads that are dense with information can be slower.
- Six rule: maximum of six lines per slide and six words per line
- Use title overheads for every significant point in the training.
- Try some unique graphics and color, but keep consistent and appropriate for the occasion.
- Use the 3T approach. Tell the students what you are going to tell them, tell them, then tell them what you told them.
- Practice, Practice, Practice—know your material and don't read the slide or your notes. Know the order of the slides, so you don't have to look at the screen to see what's next.
- Try to anticipate questions and do your homework (live fire practice is great).
- Define technical words and jargon on the first usage. Use language in your talk, which is consistent with the overheads.
- Keep charts, and text lists simple. KISS (keep it simple stupid)
- Make one point per slide.
- Check out your room, sit in several seats, and establish light levels. Can you see the slides and read the text from anywhere in the room?
- Always face your audience while speaking. Never speak into the screen.
- Make a second copy of any slide that you have to refer to again so you can insert it into the presentation in the correct sequence.
- If you have written handouts, make sure the slides are duplicated precisely.
- Prepare yourself for something to go wrong (like keeping your presentation on a memory stick or a spare laptop, carry a spare bulb for the LCD projector).

Adapted from Cinegraphics in Garden Grove, CA

Return on Investment from Training

The analysis is complete except for the vital go/no-go question—is this training worth the investment?

The returns come in two areas:

What is Calvin Smith worth as a chief technician? He has 8–10 years left that would be spent with our company (we presume but cannot guarantee). We have no guarantee that this program will work, and Calvin will learn what he has to know. Does he have the aptitude to be successful? Is his specialized knowledge built up over 12 years with us and the rest of his experience of value? Is there a training cost for his replacement? Keep in mind that servicing of microprocessor boards is only a small part of the job of Chief Technician. If we are successful in training Calvin, we might have a good prospect for an instructor to the rest of the field service staff.

We also have **a continuing asset in the training program materials that have been assembled.**

29 Dealing with and Solving Problems

Problems are the reason for maintenance departments. All maintenance professionals must become experts in the identification and solve a wide range of issues related to our assets. These issues are also a window into an opportunity. Build your muscle in problem-solving.

The person identified with formal problem solving was George Pólya. George Pólya (December 13, 1887 to September 7, 1985) was the premier teacher in problem-solving in the 1940–1960s. His approach has been taught and adopted widely.

First, you must realize you have a problem. Then you must understand the problem. He then asks, "What is the unknown?" To better see the problem, can you separate parts of the problem, can you write them down?

"Understand the problem" is often neglected as being self-evident. Pólya taught teachers how to prompt each student with appropriate questions, depending on the situation, such as:

- What are you asked to find or show?
- Can you restate the problem in your own words?
- Can you think of a picture or a diagram that might help you understand the problem?
- Is there enough information to enable you to find a solution?
- Do you understand all the words used in stating the problem?
- Do you need to ask a question to get the answer?

Second, have you seen this problem before (perhaps in a slightly different form)? Have you seen this symptom before? Was there an investigation, and is any of it useful when looking at this problem? Can you restate the question in a way that makes it more precise, can you restate it again?

The third step is to develop a plan to solve the problem. As you proceed, check your steps and make sure you made no mistakes. Can you prove that each step is correct?

There are many ways to solve problems. Skill at choosing an appropriate strategy is best learned by solving many problems.

PROBLEMS ARE GOOD!

The best organizations in the world are not the ones with all the systems and procedures, all the acronyms, or even the most up to date technical solutions; instead, they are the ones that are best at identifying and solving problems. Want more success? Want a better maintenance effort? Then get better at finding and solving problems.

I want to dissect the famous case of the sinking of the Titanic and show how problems, even horrible, tragic ones can be pointers to areas where some attention would improve the process, product, or company. Before we can delve into that, we have to look at how things go wrong.

Accidents are not accidental, and that mistakes are only as consequential as the design allows them to be.

Accidents are almost always a series of events (sometimes an improbable series) that add up to the accident. The string of events might look like this: The pedestrian was wearing black at night. It was rainy, and the tires were bald, the driver was tired, the driver was late for a delivery. At the same moment, the driver changed radio stations, the pedestrian tripped on a log and fell toward the street. What we say is that if any of those circumstances were different, avoided, or mitigated, the accident might not have happened.

What about the consequences of an event? If we blow an engine on most interstates, we can steer onto the shoulder and slow, then stop the rig. In most cases, there will be no loss of life or additional loss of property (beyond the engine).

But what if the engine powers a Piper 175 (single-engine)? Or what if the engine fails in a hospital's backup generator during a power outage? Then the consequences are dire.

Our design's multiple levels of robustness are what protect us from harm. The hospital's generator has frequent PMs and extensive annual work to make sure it starts and runs. The Piper has redundancy in the systems that could be a source of failure (ignition, fuel pumps, etc.). An airplane has what is said to be a defense in depth. Many things must go wrong for the plane to be lost. The Titanic lacked a defense in depth and suffered a series of catastrophic problems.

All companies have policies and procedures that are outmoded, not followed accurately, not understood, mistaken, and ignored. Which ones of these are dangerous? Which ones of these that, if followed, would improve the safety or effectiveness of our businesses? Having a problem and determining the causes and effects answers some of these questions.

The Titanic was 882 feet 9 inches (269.1 m) long and 92 feet 0 inches (28.0 m) abeam, the tonnage 46,328, and the height from the water line to the boat deck of 59 feet (18 m). The Titanic was the largest ship of the day. Currently, for comparison, the Oasis of the Seas, one of the largest ships, displaces almost 225,282 tons (more than four times larger).

April 14, 1912, The Titanic struck an iceberg and within hours sunk. Of a total of 2223 people aboard, only 706 survived, 1517 people perished, and the ship was lost. The total lifeboat capacity was 1178 people (which was more than the regulations called for at the time).

If we look at the causes, we can see engineering and design issues, procedure issues, lapses in judgment, and lapses in the laws and statutes that govern maritime affairs.

Of course, one complaint is that we have 20/20 hindsight. The same process could be used for any event happening today (to use the problem to improve our businesses). To see this, just think of any airplane crash and the investigation which issues directives, changes to procedures that occur. The second complaint, which is probably true, is that our advanced technology tells us things that would not have been known in 1912. Also, us knowing that the ship sank gives us an edge. The truth is that our solutions might not have even occurred to the people of that era.

At 11:40 pm, the ship hit the iceberg with a glancing blow to the starboard side, which caused buckling in the hull plates along the first five compartments. The seams of the plates popped open (only an inch or two) for almost 300 feet. This slice allowed water to rush in.

Metallurgists say that the plates were brittle at the low temperatures of the North Atlantic, and a different alloy would have been able to take the stress of the slicing blow against the iceberg. Other Metallurgists say that the rivets were substandard (rivets that were retrieved from the wreck were occluded with slag), and had they bought and installed suitable rivets correctly, the plates would have held. Let's accept them as among the causes.

Experts say that two major design flaws saved money in the manufacture. One was the walls of the watertight compartments did not extend to the ceiling. They were well above the waterline, and no one thought that would be an issue. It turned out to be pivotal in the speed of the sinking when the ship tipped up, and water flowed from a compartment to the next one.

The second piece of dicey engineering was the size of the rudder. The ship had a rudder designed for a much smaller ship. After the iceberg watch called out (they were late in seeing the iceberg) that there was an iceberg to the starboard, the ship started to turn but couldn't turn fast enough to avoid the collision. The glancing blow was much worse than a straight head-on collision (which the ship could have survived). So we have two engineering causes.

Several procedures were causes to the accident. The first one was when the iceberg was sighted, the bridge called for reversing the engines. Only two of the three propellers could be reversed. That caused turbulence and cavitation. The small rudder located in the middle of this mess was rendered even more useless by all the turbulence. Ship experts say if the engines were just slowed, the small rudder would have been adequate. The second issue is they should have turned toward the iceberg (at the last minute) to bring the stern out and around the berg.

Finally, we have the problem of seeing the iceberg too late. The ship was going like gangbusters to make its inaugural voyage a quick one. At 22 knots, it was thought that any icebergs big enough to hurt the ship would be seen well ahead of time. Two watches were in the crow's nest without binoculars. Other ships of the day used iceberg watches on the ship's bow. The Titanic captain didn't think this was necessary.

There were other causes but let's stick to these few. The crazy thing is if any of these causes had been eliminated or mitigated, the Titanic would just be another big ship from the turn of the century. It would not have become famous. It would not have sunk, or if it had, there would have been enough time for a rescue.

The tragedy caused a firestorm of publicity, and inquiries were immediately started in both the US and England. The point is that the inquiries solved problems that were there but until this incident unrecognized. These problems left unexamined and would have resulted in possibly numerous tragedies. The findings and changes to the law, design, and maritime procedures from the causes found by these inquiries have made traveling by ship an order of magnitude safer for everyone. In this way, the tragedy leads the way to a complete review of the vessel, its systems, and the whole maritime industry. Many problems were found, and most were fixed.

Some of the causes were fixed by regulation and others by interested parties alerted to the existence and nature of the problem.

Problem	Solution
Steel alloy of plate	Improved hull designs and major gains in Metallurgy solved the brittleness problem by WW1.
Waterproof compartments	Improved bulkhead design (watertight to ceiling).
Rudder size	Improved rudder standards.
Maneuver strategy	Better training, better communications.
Speed	Better regulations and rules in adverse conditions.
Watch	Make sure they have the tools they need, and they are located correctly.

Of course, not sinking is the best outcome for safety. But if you do sink what changes were made to avoid the terrible loss of life:

Lifeboats	Adequate lifeboats for all people aboard.
Better life jackets	Improved life-vest design.
Safety drills	The holding of safety drills.
Communications improvements	New policies and procedures and 24-hour radio coverage so that help will come as soon as possible.

Use these techniques the next time you have a problem (big like the Titanic or small one like a tire blew out). Uncover the causes and use them as a viewing platform to look at your whole operating situation. DuPont does this and finds an 11:1 Return on Investment from finding and fixing the causes of its problems.

My thanks to Mark Galley, Principal of Think Reliability for some of the information and approach to the Titanic sinking and to Bill Holmes, President RCA RT Melbourne, Australia.

ROOT CAUSE ANALYSIS

Anytime you have a bad situation, especially if it happens repeatedly, it may be wise to figure out what causes the condition to occur and remedy it to prevent future reoccurrences. This problem-solving technique is what's known as Root Cause Analysis, finding the roots of your problems and resolving them rather than continuing to deal with the symptoms.

General Principles of Root Cause Analysis

- The analysis must establish all known causal relationships between the root cause(s) and the defined problem.
- There is usually more than one root cause for any given problem.
- To be effective, RCA must be performed systematically, with conclusions and causes backed up by documented evidence.

- Root cause analysis can transform an old culture that reacts to problems into a new culture that solves problems before they escalate, reducing variability and improving long-term performance.
- Aiming performance improvement measures at root causes is more effective than merely treating the symptoms of a problem.

Root Cause Scene

The Plant Manager walked into the plant and found oil on the floor. He called the Foreman over and asked him why there was oil on the floor. The Foreman indicated that it was due to a leaky gasket in the pipe joint above. The Plant Manager then asked when the gasket had been replaced, and the Foreman responded that Maintenance had installed four gaskets over the past few weeks, and each one seemed to leak. Maintenance told Purchasing about the gaskets because it appeared they were all bad.

The Plant Manager then went to talk with Purchasing about the situation with the gaskets. The Purchasing Manager indicated that they had received a bad batch of gaskets from the supplier. They had been trying for the past two months to try to get the supplier to make good on the last order of 5000 gaskets that all seemed to be wrong. The Plant Manager asked the Purchasing Manager why they had purchased from this supplier if they were so disreputable; he answered that they were the lowest bidder when quotes were received from various suppliers.

The Plant Manager asked the Purchasing Manager why they went with the lowest bidder, and he indicated that was the direction he had received from the VP of Finance. The Plant Manager then went to talk to the VP of Finance about the situation.

When the Plant Manager asked the VP of Finance why Purchasing had always been directed to take the lowest bidder, the VP of Finance said, "Because you indicated that we had to be as cost-conscious as possible!" and purchasing from the lowest bidder saved lots of money.

Bingo!

The Plant Manager was horrified to realize that he was the reason there was oil on the plant floor.

COMMON METHODS OF RCA

Cause-and-effect analysis: A technique that organizes the analyst's knowledge into a cause-and-effect chain. For every effect, there is a cause. There often is a long string of relationships between the causes and their consequences. In theory, if the lowest cause on the chain is removed, the problem will not re-appear. By establishing an investigatory mindset, an organization can transform itself from an old culture that reacts to problems to a new culture that solves problems before they escalate.

Failure Mode and Effects Analysis (FMEA) "A failure modes and effects analysis (FMEA) is a procedure for analysis of potential failure modes within a system for classification by the severity and likelihood of the failures. A successful FMEA activity helps a team to identify potential failure modes based on experience with similar products or processes, enabling the team to design those failures out of the system with the minimum of effort and resource expenditure, thereby reducing development time and costs." (Wikipedia)

Pareto analysis: Pareto analysis is a formal technique useful where many possible courses of action compete for your attention. The process counts the incidents and assigns a probability based on the counts. The Pareto analysis concerns itself with the 20% of "bad actors" that cause 80% of the problems.

Pareto analysis is a creative way of looking at causes of problems because it helps stimulate thinking and organize thoughts. However, it can be limited by its exclusion of possibly essential issues, which may be small initially, but which grow with time. It should be combined with other analytical tools such as FMEA and Fault Tree analysis, for example.

Ishikawa Diagram: also known as the fishbone diagram or cause and effect diagram, looks at the contribution of the person, the materials, the machine, and the method to the problem. The fishbone diagram was developed for manufacturing and is widely used in that context.

DETERMINE THE ROOT CAUSE: FIVE WHYS

Asking "Why?" may be a favorite technique of a three-year-old child, which drives everyone else crazy, but it could teach you much valuable information.

By repeatedly asking the question "Why" (five times is a good rule of thumb), you can peel away the layers of symptoms, which can lead to understanding the root cause of a problem. Very often, the ostensible reason for a problem will lead you to another question. Although this technique is called "five Whys," you may find that you will need to ask the question fewer or more times than five before you find the issue related to a problem.

Benefits of the five whys

- They help identify the root cause of a problem.
- They determine the relationship between different root causes of a problem.
- Five Whys is one of the most straightforward tools, easy to complete without statistical analysis.

How to complete the five whys?

- Write down the specific problem. Writing out the issue helps formalize the problem and describe it completely. It also helps a team focus on the same problem.
- Ask why the problem happens and write the answer down below the problem.
- If the answer you just provided doesn't identify the root cause of the problem that you wrote down in step 1, ask **Why** again and write that answer down.
- Loopback to step 3 until the team is in agreement that the problem's root cause is identified. Again, this may take fewer or more times than five Whys.

Bill Wilson points out some shortcomings to the five Why approach. Once you have a plausible root cause, he suggests asking these additional questions:

- What proof do I have that this cause exists? (Is it concrete? Is it measurable?)
- What proof do I have that this cause could lead to the stated effect? (Am I merely asserting causation?)
- What proof do I have that this cause contributed to the problem I'm trying to fix. (Even given that it exists and could lead to this problem, how do I know it wasn't something else?)
- Is anything else needed, along with this cause, for the stated effect to occur? (Is it self-sufficient? Is something needed to help it along?)
- Can anything else, besides this cause, lead to the stated effect? (Are there alternative explanations that fit better? What other risks are there?)

Checklist for Possible Root Causes

Materials

Defective raw material
Wrong type for the job
Lack of raw material

Machine/Equipment

Incorrect tool selection
Poor maintenance or design
Poor equipment or tool placement
Defective equipment or tool

Environment

Disorderly workplace
Poor job design or layout of work
Surfaces poorly maintained
Physical demands of the task
Forces of nature

Management

No or poor management involvement
Inattention to task
Task hazards not guarded properly
Other (horseplay, inattention)
Stress
Lack of standardized process

Methods

- No or poor procedures
- Practices are not the same as written procedures
- Poor communication

Management system

- Training or education lacking
- Poor employee involvement
- Poor recognition of the hazard
- Previously identified hazards were not eliminated

30 Continuous Improvement

Continuous improvement is the application of problem-solving continuously. The result is a better environment now than last year and a better environment next year than today. It is no less than one of the main features of survival and thriving of an organization

DEFECT ELIMINATION

Why Eliminate Defects?

- Less maintenance work especially breakdowns
- Better quality products
- Better yield from raw materials
- Improved Uptime
- Higher reliability with fewer machine breakdowns
- Longer equipment life
- Higher energy efficiency
- Fun and develop new skills

Here is the bad news. You have a giant pool of defects ready, at any moment, to poison your maintenance and operation.

Here is the worse news. Defects are flowing in faster than you are removing them!

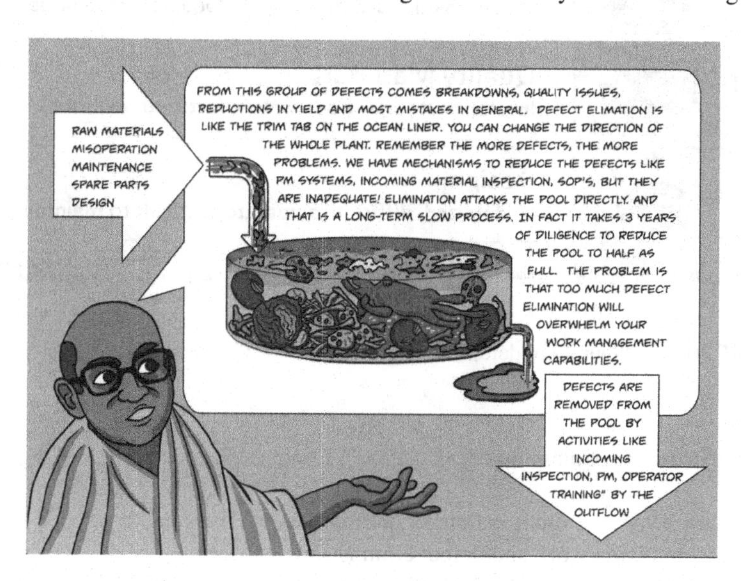

FIGURE 30.1 Guru explaining the impact of defect elimination.

The 1% solution to defect elimination (work by Winston Ledet):

- Scientific modeling shows that a useful metric is 1 out of every 100 corrective work orders be "Don't Just Fix It! Improve It!" work order.
- More than 1% overwhelm the work execution management system.
- If you stick to 1% defect elimination, in three years, 54% defects eliminated.
- 73%+ defects eliminated in six years.

What Is a Defect?

- Anything that erodes value reduces production, compromises HSE, or creates waste.
- Fewer defects mean less work, better product, higher production, and a safer workplace.
- Everywhere you can imagine. And in every activity, you do.

FIGURE 30.2 5 Sources of defects.

Focus on action, not planning

- If reactive—the focus is first on breakdowns.
- Use precision work and defect elimination.
- Engaging everyone to put equipment back into production "better than new."

(don't just fix it, improve it).

- Teams with diversified members who will act together.
- Focuses on results (projects complete) in a short period.
- Promotes learning of new skills and behaviors.
- Disbands or chooses to remove or prevent another defect.

Our primary battle is against: "Random acts of Carelessness."
–Gabriel Delgado, Freeport McMorran

31 Using Projects to Motivate the Team

Small autonomous projects for defect elimination, lean maintenance, or other improvement are an excellent method to both motivate your team and to improve the overall maintenance situation.

Projects are the smallest activities that work toward the realization of some of the bigger goals of the workgroup. Projects come in all sizes. In this discussion, we are limited to DE and smaller lean projects.

Type of project	~Cost range	Management involvement	Engineering	MOC	Initiating reason	Reporting
IYF	0-$25	None	Informal or none	None	IYF!	None or log
DE project	0-$2500	Little or none	By DE person or team	None or Little	Defect-easy first	A short note or closing comment
Lean project	$500-5,000	One person	By lean team	None, light review	Fulfillment of Aim	Short report
Improve ment project	$1000-$25,000	Two departments	By Maint. and operations	MOC as needed	Improve maintenance or operations	Report
CAPEX project	$5000 up	Full	Full engineering	Full MOC	ROI	Full report

FIGURE 31.1 Improvement projects by size and attributes.

What projects accomplish

- Forward the action toward your organization's mission, Vision, and values.
- Improve reliability in a concrete way (such as eliminating defects).
- Transform culture (create good turbulence by attacking errors and mistakes).
- Improve the working conditions of the organization.

- Have fun developing new skills and activities.
- Allow you to develop as a leader, manager, and team player.
- Use more of your talent.

HOW GOALS IMBEDDED IN PROJECTS MOTIVATE

A **goal** is a desired result a person envisions, plans, and commits to achieve a personal or organizational desired endpoint in some sort of assumed development.

To see the mechanism, we must peel back the covering. Goals have potent effects on people. Goals call people to be in action, and goals can make everyday activities more fun by creating a game.

Projects to eliminate defects

- Deep alignment to the mission and vision of the organization.
- Projects always use active language.
- Eliminates the source of the problem when possible and reduces waste.
- High Probability of Success.
- Least (no) money invested.
- Tools and materials readily available.
- Doable within 90 days.
- Shortest cycle for payback.
- Special case: Increases knowledge, a good project might be successful even if it has a negative result.

The first order of business is to go after Low Hanging Fruit! Get the easy stuff!
Assignment: DE management project

- Devote about 30–60 minutes a week to defect elimination.
- Focus on the obvious ones that lend themselves to simple solutions.
- Set-up a work order category for DE.
- Charge all labor, materials to work order.
- OK to be members of different teams and work on various defects.
- Be alert for too much DE!

Close Relative of Defect Elimination Is Lean Maintenance

Focus is on waste which is mostly a subset of defects

But the focus is different. The focus is on "Muda." The seven wastes define muda.

Muda: (無駄) is a Japanese word meaning "futility; uselessness; wastefulness," and is a crucial concept in lean process thinking, like the Toyota Production System (TPS) as one of the three types of deviation from optimal allocation of resources (the others being mura and muri).

The seven wastes consist of:

- Overproduction: Overproduction is to manufacture an item before it is required.
- Waiting: Whenever goods are not moving or being processed, the waste of waiting occurs.
- Transporting.
- Inappropriate Processing.
- Unnecessary Inventory.
- Unnecessary/Excess Motion.
- Defects.

Section IV

The Future of Managing Maintenance

32 Making Technology, Your Friend

Maintenance uses tools. Many of the new tools are changing from steel to silicon. These high-tech tools are advancing rapidly. By nature, maintenance folks adopt changes slowly (if it works, don't mess with it). Consumer products are worlds ahead of those we use. High-Tech is catching up with our field. Make it your friend—BAE (Become an Expert).

It seems society is going somewhere at an incredible pace. Today consumer products and applications are blazing a path with high tech products like Seri and Alexa. They use big data, AI, cloud, and just about everything else. Ads are popping up about the cloud, AI (artificial intelligence like IBM's Watson), big data, and machine learning.

In the present, most people imagine the change to be more comfortable than it will in fact will be. Afterward, we also underestimate the depth and pervasiveness of the change. In the early stages, there will be tons of fluff and ideas that don't go anywhere. But I suggest you hang onto your seat belt because we are going whether you or I like it or not!

There is a truth in manufacturing as well as everywhere else. Peggy Liao, Senior Director, Marketing Communications of Decisive, described it as getting the right information into the right hands, in time to do something about it.

Her example was if an alternator takes three hours to change, and the unit is in the bay, the daily report is too late to intervene. You have no opportunity to process the recalls that are sitting on your desk. But if you got a text while the machine was still there, you could intervene in time to make a difference.

HOW CAN ALL THIS NEW TECH AFFECT YOU?

As the CMMS, on-board sensors, dispatch and order entry, and real-time GPS data start to be accessible; decisions will be improved. In some cases, the maintenance advantage will be small, but the operational advantage will be significant. Other instances in which avoiding a breakdown by the timely intervention (based on sensors transmitting real-time to the cloud) could have considerable maintenance advantage.

A few of the more straightforward maintenance improvement ideas include:

- You can schedule PMs when they will impact the production cycle the least. That improves the available hours.
- Shop scheduling will be more reliable with fewer disruptions.
- Catastrophic events will fall dramatically because some of those events broadcast themselves from the sensors (like oil, water problems).

- Rebuilds will be better scheduled; you'll replace only the worn parts.
- The list can go on, but the idea is better information in the hands of the people who can use it supports intelligent decision making.

SUPERVISOR'S GUIDE TO PREDICTIVE MAINTENANCE TECHNIQUES

The human sense inspection activity on the PM task list is predictive (if you use the dictionary definition). In this chapter, we make the distinction between human sense inspection and techniques that require special tools, instruments, or materials.

Predictive Maintenance is a maintenance activity geared to indicating where a piece of equipment is on the critical wear curve (P/F curve, the earlier, the better) and giving you a rough idea of its useful life. The instruments generally amplify (vibration) or change the frequency (Infrared) of the signal so we can detect it.

Maintenance has borrowed tools from other fields such as medicine, chemistry, physics, auto racing, aerospace, and others. These advanced technology techniques include all types of oil analysis, chemical analysis, infrared temperature scanning, Magna-flux, ultrasonic imaging, shock pulse meters, and advanced visual inspections.

All techniques compare the reading of the instrument to a known engineering limit. These technologies can help the supervisor's team detect problems with equipment well before it becomes a problem. Technology has been improving significantly in this area. Tools are available that can investigate an air handler, thread through boiler tubes, or detect a bearing failure, weeks before it happens.

All the predictive techniques (except condition-based maintenance) we are going to discuss should be on the PM task list and controlled by the PM system.

Predictive maintenance is partly an attitude, not necessarily a new technology. Other instruments not discussed, such as meggers, pyrometers, strain gauges, and temperature-sensitive tapes and chalk, used correctly, can also be useful in a predictive way.

PdM: Consider These Questions

1. Is the specific technique, the right one?
    - Does the return justify the extra expense?
    - Do you have existing information systems to handle and store the reports?
    - Is it easy and convenient to integrate the predictive activity with the rest of the PM system?
    - Is there a less costly technique to get the same information?
    - Will the technique minimize interference to our users?
    - Exactly what critical wear are we trying to locate?
2. Is this the right vendor?
    - Will they train you and your staff?
    - Do they have an existing relationship with your organization?
    - Is the equivalent equipment available elsewhere?
    - In the case of a service company, are they accurate?

- How do their prices compare with the value received in the marketplace?
- In most metropolitan locations, service companies are available to perform these services.

3. Do you have the support to make this effective?
   - Commitment for a year or more.
   - Executive sponsor.
   - Budget for training a couple of people in-depth.
   - Budget for basic training for a wider group.
   - Engineering support.
   - Time to build expertise.

Lots of technologies

Oil analysis
Vibration analysis
Ultrasonic
Temperature
Infrared
Eddy current testing
Magnetic Particle
Motor current analysis
Megger testing
Special optical
NDT
Other

You Only Need a Few Technologies to Detect Problems

Your choices depend on the equipment you use and the expertise of your team. Just like in medicine, you might consider using different techniques to uncover or verify findings.

Oil Analysis

One of the most popular techniques to predict current internal conditions and impending failures is oil analysis. It is best used for all types of engines, turbines, large gearboxes. Other forms of analysis include water analysis and dielectric testing.

Oil analysis is shown to be essential because of a study of drums of new oil. The latest research showed that new drums of oil are more contaminated then allowed by the specification. It behooves you to have new oil tested before usage. It is probably a good idea to have the new oil verified by an independent service. Drums should not be stored outdoors since they expand and contract (breath in water and contamination) with changes in temperature.

Spectrographic techniques are commonly used to analyze the contents of the oil. Spectrometry report all metals and contamination since different materials give off different characteristic spectra when burned. The results are expressed in PPT or PPM (parts per thousand or parts per million).

The lab or oil vendor usually has baseline data for the types of equipment they frequently analyze. The concept is to track trace materials over time and determine the source. At a level, the experience will dictate an intervention is required (a re-build or re-manufacture). Oil analyses cost $10–$25 (and up for specialized applications) per report. Oil analysis is frequently included at no charge (or low charge) from your supplier of oil.

You are usually given an emailed report with a reading of all the materials in the oil and the "normal" readings for those materials. In some cases, the lab might call the results so that you can stop a unit to avoid more damage.

For example, if the level of silicon is high in the oil, most likely, a breach has occurred between the outside world and the lubricating systems (frequently silicon contamination comes from sand or dirt). Another example would be an increase from 4 PPT to 6 PPT for bronze, which probably indicates increasing bearing wear. This reading would be tracked and could be noted and checked on the regular inspections.

Oil analyses include an analysis of the suspended or dissolved non-oil materials including babbitt, chromium, copper, iron, lead, tin, aluminum, cadmium, molybdenum, nickel, silicon, silver, and titanium. In addition to these materials, the analysis will show contamination from acids, bacteria, fuel, water, plastic, and even leather.

The other aspect of oil analysis is a view of the oil itself. Questions answered include: has the oil broken down, what is the viscosity, are the additives still available? Other tests are carried out on power transformer oil, which shows dielectric, and breakdown properties (a major transformer outage could disrupt your whole facility!).

Consider oil analysis as a part of your normal PM cycle. Since oil analysis is relatively inexpensive, you should also consider doing it:

1. Following any overload or unusual stress
2. If sabotage is suspected
3. Just after (or better yet before) purchasing a used unit
4. After a bulk delivery to determine lube quality, and especially if bacteria is present
5. Following a rebuild, to baseline the new equipment and for quality assurance
6. After service with severe weather such as floods, hurricanes, or sandstorm.

Vendors

Keep in mind the best place to begin looking at oil analysis is from your lubricant vendor. If your local distributor is not aware of any programs, contact any of the major oil companies. If you are a large oil user and are shopping for a yearly requirement, you might ask for analysis as part of the service. Some vendors will give analysis services to their more substantial customers at little or no cost. Unaffiliated labs exist in most major cities, especially cities that serve as manufacturing or transportation centers. These firms will prepare a report of all the attributes of your hydraulic, engine, cutting oils, or power transmission lubricants.

Tip: Send samples taken at the same time on the same unit to several oil analysis labs. See who agrees, who is the fastest, who has the least cost.

Vibration Analysis

Vibration Analysis is a widely used method in plant/machinery maintenance. A study of the city of Houston's wastewater treatment department showed a $3.50 return on investment for every $1.00 spent on vibration monitoring. Their research showed that a private company might get as much as $5.00 profit per dollar spent. The study and the vibration-monitoring project were done by the engineering firm of Turner, Collie, and Braden of Houston, Texas.

Three types of Vibration Measurement:

- Displacement
- Velocity
- Acceleration

Vibration analysis measures the changes in the amplitude of the vibration by frequency over time. This amplitude by frequency is plotted on an XY axis chart and is called a signature (for a given service load). Changes to the vibration signature of a unit mean that a vibration element has changed characteristics. Vibration elements include all rotating parts such as shafts, bearings, motors, power transmission components. Also included are anchors, resonating structures, and indirectly connected equipment.

Ultrasonic Inspection

One of the most exciting technologies is ultrasonic inspection. It was first used in medicine (obstetrics) and moved to factory inspection and maintenance. In medicine, an ultrasonic transducer transmits high-frequency sound waves, and a specialized microphone picks up the echo. Echoes are caused by changes in the density of the material tested. The echoes are assembled by computer into pictures for the doctor.

Ultrasonic gauges use a simplified technique of timing the bounce of the ultrasonic signal and can determine the thickness of metal, piping, corrosion of almost any homogenous material. One exciting application is the shock-pulse meter, which reads the film thickness of oil on bearings.

The most prominent use of ultrasound in maintenance is a different application is in the area of ultrasonic detection. Many flows, leaks, bearing noises, air infiltration, and mechanical systems give off ultrasonic sound waves. Portable detectors can quickly locate the source of these noises and increase the efficiency of the diagnosis. Here the advantage of high-frequency waves is that they are highly directional. An Ultrasound gun can "hear" a compressed air leak from 30' and point directly at the leak.

Uses of ultrasound:

- Compressed air leaks
- Vacuum leaks
- Corona discharge

- Motor faults
- Lubrication and bearing condition
- Gearbox condition
- Steam trap inspection

Low Hanging Fruit in Many Plants: Ultrasonic Compressed Air Survey

With the system on, follow all the compressed air pipes from the compressors to the points of use.

Small leaks whistle in ultrasonic frequencies. You can hear the whistle (which is highly directional) with the ultrasonic microphone. Tag the leak. Ambient noise is not a problem.

Savings is in electricity is immediate. Sometimes the reduction in the need for compressed air is enough that an additional compressor is not needed. If that is so, you can reinstate one of your compressors as an actual back-up so that one can be taken down for a no-downtime service (as in the original design).

When you can, shut down the system fix the leaks. Re-survey plant periodically or when there has been a significant change.

TEMPERATURE MEASUREMENT

Since the beginning of the industrial age, temperature sensing has been an important issue. Friction (or in the electrical field, resistance) creates heat. Temperature is the single greatest enemy for lube oils and the power transmission components. Advanced technologies in detection, imaging, and chemistry allow us to use temperature as a diagnostic tool.

Today, we have the technology to photograph by heat rather than reflected light—the camera's use of false color. The camera shows hotter parts as white to redder (to blue for cool parts). Changes in heat will graphically display problem areas where wear is taking place or where there is excessive resistance in an electrical circuit of friction in a mechanical system.

Readings are taken as part of the PM routine and tracked over time. Failures show up as an increase in temperature.

Many insurance companies require annual infrared surveys of all electrical distribution. Insurers know hot spots become fires and claims.

Temperature detection utilizes infrared scanning (video technology), still film, pyrometers, thermocouples, other transducers, and heat-sensitive tapes and chalks.

On larger stationary engines, air handlers, boilers, turbines, etc., temperature transducers measure all significant bearings. Some packages include shutdown circuits and alarms if the temperature gets above certain limits.

Possible uses for infrared inspection	To Look for
Bearings	Overheating
Boilers	wall deterioration
Cutting tool	Sharpness
Die casting/injection molding equipment	temperature distribution

Distribution panels	Overheating
Dust atmospheres (coal, sawdust)	spontaneous combustion indications
Furnace tubes	heating patterns
Heat exchange	proper operation
Kilns and furnaces	refractory breakdown
Motors	hot bearings
Paper processing	uneven drying
Piping	locating underground leaks
Polluted waters	sources of dumping in rivers
Power transmission equipment	bad connections
Power factor capacitors	Overheating
Presses	mechanical wear
Steam lines	clogs or leaks
Switchgear, breakers	loose or corroded connections
Three phase equipment	unbalanced load
Thermal sealing, welding, induction heating	even heating

Examples of areas where savings are possible from the application of infrared:

- A hot spot on the transformer was detected. The repair was scheduled off shift when the load was not needed avoiding costly and disruptive downtime.
- A percentage of new steam traps, which remove air or condensate from steam lines, will clog or fail in the first year.
- Hot bearings were isolated in a production line before deterioration had taken place.
- Roofs with water under the membrane retain heat after the sun goes down.
- Furnaces are excellent places to apply infrared because of the cost involved in creating the heat and the cost of keeping in place.
- Temperature measurement in electrical closets

Advanced Visual Techniques

The first applications of advanced visual technology were using fiber optics. With fiber optics, flexible fibers of highly pure glass are bundled together. Each strand of glass carries a small part of the picture. The most miniature fiber optic instruments have diameters of 0.9 mm (0.035″). Some of the devices can articulate to inspect the walls of a boiler tube. The focus on some of the advanced models is 1/3″ to infinity. The limitation of fiber optics is the length. Generally, they go no longer than about 6 feet. The advantages are cost (about 50% or less of equivalent video technology and level of technology (they do not require large amounts of training to support).

Another visual technology gaining acceptance is ultra-small video cameras. CCD cameras inspect the interior of large equipment, boiler tubes, and pipelines. This CCD (Charge Coupled Display) device can be attached to a color monitor through cables or Wi-Fi. They use a miniature television camera smaller than a pencil (about 1/4″ in diameter and 1″ long) with a built-in light source. Some models allow small tools to be manipulated in the end; others can snake around obstacles. They are extensively used to inspect pipes and boiler tubes.

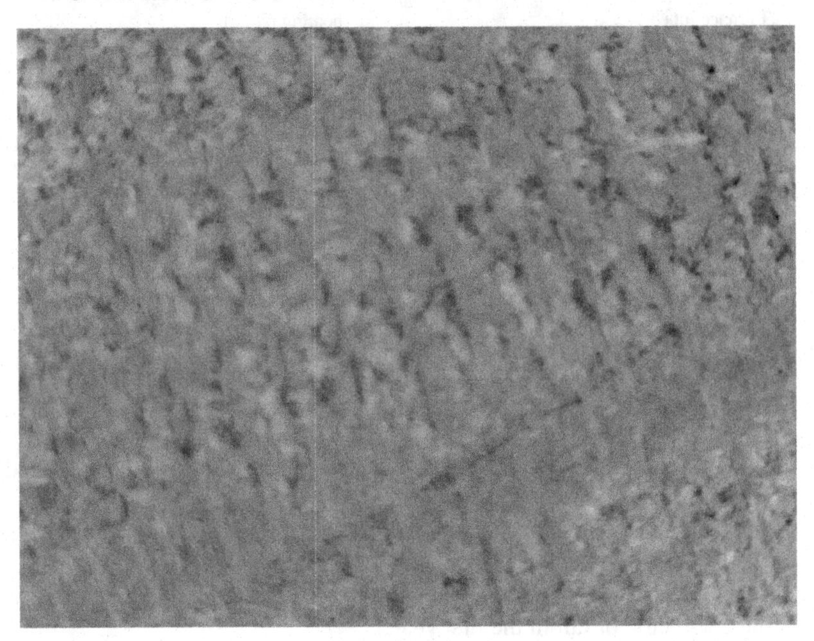

FIGURE 32.1 Picture of wood from a $30 microscope.

This technique has become popular for detecting and locating sewer problems and can save thousands of dollars in digging.

In most major industrial centers, service companies can do your inspections for a fee. These firms use the latest technology and have highly skilled inspectors. Some of these firms also sell hardware with training. One suitable method is to try some service companies and settle on one to do inspections, help you choose equipment, and do training.

A Few Other Methods of Predictive Maintenance

Magnetic Particle Techniques (called Magna-flux or eddy current testing)

Magna-flux is borrowed from racing and racing engine rebuilding and is used in general maintenance. This technique induces very high currents into a steel part (typically used on shafts like crank and camshafts). While the current is being applied, the piece is washed by fine, dark-colored magnetic particles (there are both dry and wet systems).

The test shows cracks that are ordinarily too small to be seen by the naked eye and cracks that end below the surface of the material. Magnetic fields change around cracks, and the particles outline the areas. The test was initially used when re-building racing engines (to avoid putting a cracked crankshaft back into the engine). The high cost of parts and failure can frequently justify the test. The OEM's who built the cranks and cams also use the test as part of their quality assurance process.

Penetrating Dye Testing

Penetrating dye testing is similar visually to Magna-flux. The dye gets drawn into cracks in welded, machined, or fabricated parts. The process was developed to inspect welds. The penetrating dye is drawn into cracks by capillary action. Only cracks that come to the surface are highlighted by this method.

IIoT Sensors (Industrial Internet of Things)

The predictive maintenance techniques already discussed have been around for 40 or more years. The latest trend is putting sensors directly on the equipment to transmit a reading over WIFI, Bluetooth, wires to programs that can analyze the data stream.

Automation of Conditional-Based Maintenance

Some Background

Condition-based maintenance has always been a desirable maintenance PM option. A maintenance action is condition-based when gauge readings or condition initiates it—changing a filter when the differential pressure from the gauges before and after the filter reached a certain level.

The issue was the cost of collecting the data to make the condition-based decision. Truck fleets have been doing this for ages. The drivers would always bring their trucks in when the oil light or brake lights went on.

The solution to CBM was on-line sensors that you did not have to read. On-line and wired sensors have been available for decades. There were hundreds of types used by the process industry. The challenge to widespread use was costly installation and wiring. Frequently the wiring was 90% or more of the cost of the project.

Smartphones and other consumer products stimulated the development of smaller and less expensive sensors. Once removed from the phones and given more robust packaging, these new sensors were ultralow users of power. They could run (at this point 10 years and some even use micro-generators that never need batteries) years on a single battery.

IIoT Sensors

Almost every technology for predictive maintenance got turned into these

- RFID
- Motion
- Cameras
- Hi-speed capture
- Temperature
  - Duct temp
  - Water temp
  - RTD high, low

- Vibration (accelerometers)
- Sound level
- Mechanical stress
- Vibration
- Dry contact
- Open/closed
- Water present
- Voltage
    - 0–5 V
    - Voltmeter
    - Volt detection
    - Conductivity
- Pressure
- Spectrometer
- Resistance
- Amps
    - Amp meter
    - Current detect
- Tilt
- Impact
- Movement
- G force (snapshot)
- Weather
- Colorimeter
- Turbidity
- CO_2
- PH
- Light

HIGH TECH INVASION

High tech has invaded every aspect of our lives. In fact, without it, our society might very well collapse. Maintenance management is rather late to this party. Advances in sensors, massive cloud storage, fat (high-speed, wide-bandwidth) communication data pipes, analytics, and AI (Artificial Intelligence) came first to consumer goods, then to production processes, and now finally to maintenance.

We mentioned the sensors in the prior section. The sensor business is over 75 years old and was essential for any process plant. The scope of sensing has expanded because of the drive for miniaturized and low-cost sensors for smartphones, wearables, and other consumer gadgets.

The range and variety of data are now vast. It extends from traditional sensors to GPS data, production data, scanned written records, weather data, and so on. They are called "Big data" when you add in data streams from other sources. We could add in weather, incoming orders, traffic conditions, fuel prices along the route, and almost anything else that could impact our decisions.

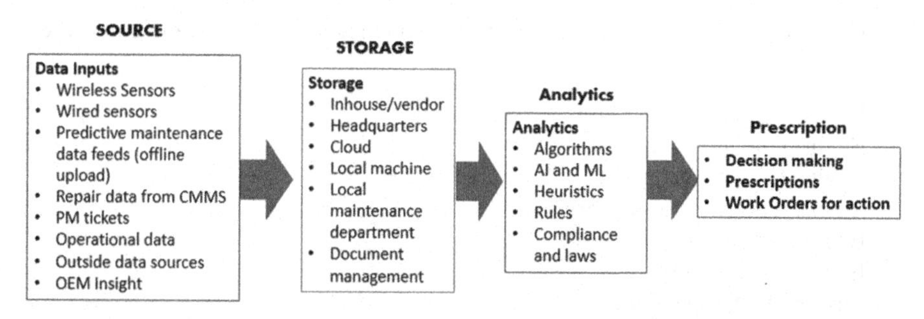

FIGURE 32.2 Block diagram for high tech maintenance.

All this data travels through fat data pipes (meaning wide bandwidth wires that can carry tons of data) to the cloud storage. The cloud is simply a rented server and storage space.

Once everything (all the data) is in one place, advanced analytics can slice and dice the data hundreds or thousands of ways. We can also use Artificial Intelligence to review all the data streams to help us plan for the future and make optimized decisions in the present.

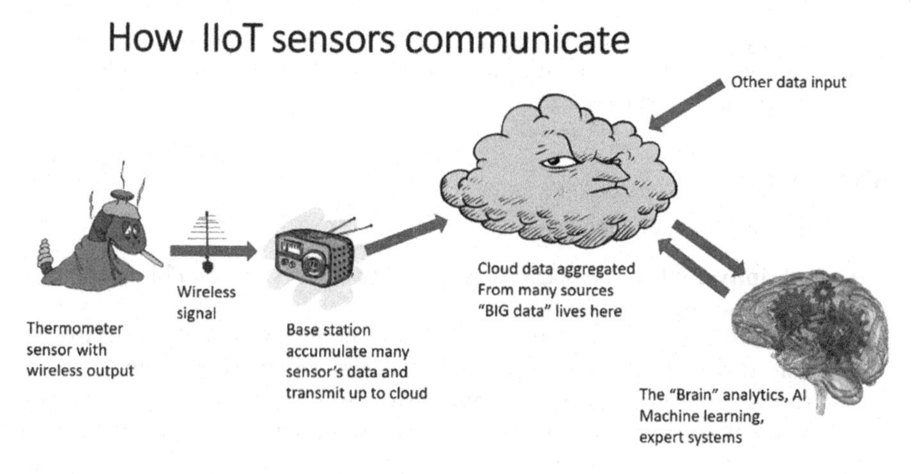

FIGURE 32.3 Flow diagram of sensor to computer information.

Example of Modern IIoT

For example, a Caterpillar mining haul truck has 145 sensors. Each sensor reports its readings every second. The data files start to get quite large and are sent to a local cloud.

Here are a few of the 145 sensors that are putting readings into the cloud every second.

Engine Coolant Pump Outlet Pressure
Engine Coolant Temperature
Engine Coolant Pump Outlet Temperature
Engine Coolant Temperature
Engine Oil Level
Engine Oil Pressure
Engine Oil Temperature

Ground Speed
Fuel Consumption Rate
Fuel Filter Differential Pressure
Fuel Pressure
Fuel Rail Pressure
Fuel Rail Temperature
Fuel Temperature
Fuel/Water Separator Level Status

Some of you are experts in machine engine operation and maintenance. If you looked at the numbers from each sensor and their relationship with each other, relation to the load out data, relationship to the weather, and changes in altitude (from hills and valleys), what could you see? Let us add-in that you don't have to sleep, take time off, eat, or take breaks. And by the way, you can work 24/7!

Could you see a lot about how the next few days or weeks will go? I can't imagine what defects you (an engine/haul machine expert) could detect well before failure or what conditions could a simple adjustment make a world of difference to efficiency. That is the essence of analytics

ANALYTICS

There are four flavors of analytics.

- **Descriptive:** What is happening like Number of work orders, PM incomplete, Age of backlog.
- **Diagnostic:** Why is this happening? Tire problems went up because we changed vendors, Overtime is increased because breakdowns are up.
- **Predictive:** What is likely to happen (We always wanted this!) The bearing will last until Thursday, in two hours a fire will start in E-123, We will be short two electricians next year.
- **Prescriptive:** What is the best course of action? Change the darn bearing! Increase the stock level by four units to achieve a 97% service level.

A simple analysis would tell us:

- A very accurate way to predict when to change the fuel filter (Fuel Filter Differential Pressure)

- Change oil might be indicated by (Engine Oil Level, Engine Oil Pressure, Engine Oil Temperature, mileage, viscosity)
- MPG Efficiency might be related (Ground Speed, Fuel Consumption Rate)

More sophisticated analysis might give us insight into:

- Operating parameters that optimize fuel use
- Conditions that predetermine expensive failures like premature engine or transmission failures
- Specific conditions right before a major fault

All these analyses improve the ROI (return on investment), improve unit life, improve delivery integrity.

Fancy Word for Simple Idea: Algorithm

- Detailed instructions on how to solve a problem.
- The algorithm is an unambiguous specification of how to solve a class of problems.
- Algorithms can perform calculations, data processing, and automated reasoning tasks.
- A process or set of rules to be followed in calculations or other problem-solving operations, especially by a computer.

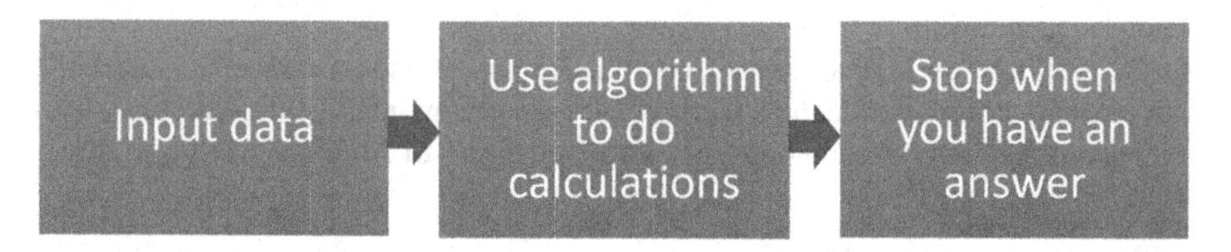

FIGURE 32.4 Algorithm.

There are thousands (or more) of algorithms.

Some have names like:

- Decision trees (a series of decisions or gates)
- Bayesian with increasingly accurate estimates
- Linear regression
- Ordinary Least Squares Regression
- Clustering

These algorithms are the business of data scientists. Data science is a relatively new field that helps us (we are called subject matter experts) deal with the data to do analytics to get us the answers to the questions we need. Some items might be, are any units going to fail in the next 48 hours. Another related question might be, are there any units we can intervene on that will avoid failure. Many questions are useful. At some point, you might work with a data scientist.

On a side note, I believe this analytics stuff is currently missing from most of our uses of CMMS. CMMS has a lot of analytics already built-in. Way more could be done by using those capabilities today, and not more investment!

ARTIFICIAL INTELLIGENCE (AI)

The conditions are right for useful AI in the management of maintenance of buildings, plants, and mobile equipment. We increasingly have the data from sensors, and elsewhere, the cloud is ready and waiting, and lots of people are developing algorithms for analysis of maintenance big data.

- IIoT Sensors
- Manual entry sensors
- Other data
- Big data
- Cloud storage
- Analytics with algorithms
- Potential AI

Artificial—made or produced by human beings rather than occurring naturally, typically as a copy of something natural. **Intelligence**—the ability to acquire and apply knowledge and skills

By the way, Seri, Alexa, Waymo, Netflix, Cortana, Watson, Google, and Facebook advertising, all use AI. Machine learning, deep learning, intelligent agent, bots, neural networks, robots, autonomous driving are all forms of AI or use AI to operate.

AI Vocabulary Cheat Sheet

- **Artificial intelligence:** AI is the development of computers capable of tasks that typically require human intelligence. Traditionally AI had to be able to pass the Turing test (Turing was a mathematician that lead the team to break the Enemiga code in World War II). The test is to interact with a computer program. If a person thought they were interacting with another person, the program passed the Turing test and was considered AI.
- **Machine learning:** Using example data or experience to refine how computers make predictions or perform a task.
- **Deep learning:** A machine learning technique in which data is filtered through self-adjusting networks of math loosely inspired by neurons in the brain.
- **Supervised learning:** Showing software labeled example data, such as photographs, to teach a computer what to do.

- **Unsupervised learning:** Learning without annotated examples, just from the experience of data or the world—trivial for humans but not generally practical for machines. Yet.

AI is a series of Algorithms that can do a variety of things. Like, use the data to come up with actions. Similar actions then perhaps an experienced maintenance person would come up with if they had the time, attention, and memory to look at the data.

Think of a bunch of algorithms strung together, a bunch of logic and mathematics, and when you have enough, you'll start to have artificial intelligence!

33 Going Forward

Only a small percentage of people think through their career where they want to advance and develop. While many people will go after training in their core areas, it is essential to improve their knowledge outside their core expertise (accounting, engineering, business, etc.).

CAREER DEVELOPMENT

It is not a mysterious process. It is a process where you ask a fundamental question about what you would like to be doing in your life. Once you've answered this question, you can answer the question of your development (what skills, experience, or qualifications you need to reach your goals).

The exciting part about this process is that you don't have to be right for the method to be effective. The goal setting is the hard part. This process puts you in good company because only 3% of the working population has written goals (guess who gets what they want more frequently, the 97% who don't set goals or the 3% who do).

The model I would like to introduce is that career development is like a savings account. In a savings account, you save slowly for distant goals ($10 a week will fund a child's education if you start young enough). You save more quickly for shorter-term goals ($200 a month for Christmas from July). In career development, short-term goals should take up most of your effort. The power of this method is that it hands you a road map for your career.

Examples, for a New Supervisor

One-year goals: Become comfortable with a new job, learn more psychology to motivate workgroup, and make a good impression on the maintenance manager.

Suggestions on things to do to achieve goals: Talk to older supervisors about what it was like for them, read one book on supervision, keep a notebook about observations of what motivates people, look for one thing in my area where a small investment will bring good returns and present it to the boss.

Five-year goals: Increase visibility in the organization, be a candidate for Assistant Maintenance Manager.

Suggestions on things to do to achieve goals: Join the local chapter of a trade association (and toward the two-three year mark become active); take a night school course in writing; take a night school course in accounting; look for larger areas of money savings; write a formal proposal for the savings; read two books a year on maintenance management; look for opportunities to expand your understanding of your field (such as taking a day during a vacation in another city to visit a colleague and tour their facility).

Ten-Year Goals: Be a Maintenance Manager

Suggestions on things to do to achieve goals: Become active in the national trade organization; complete a degree in business, engineering, join AMA (American Management Association) and attend several seminars; read two books per year on management, your industry, or maintenance; look for opportunities for leadership (outside work such as PTA, civic, church organizations); write a trade magazine article on something you know about (trade journals are usually easy to get published in if your material is timely).

HOW TO IMPROVE YOUR MAINTENANCE IQ

Nothing changes; nothing improves until you do something different from what you did yesterday. Your maintenance IQ will only gradually improve unless you take a proactive approach toward your development.

Keys to Your Development

- Keep an open mind. People are often trapped by what they think they know, by things that they didn't consider at all or never heard of.
- Decide or investigate what competencies are needed for your present job and for the job you would like to have next. Look at your broader life. What do you care about? What else do you need to be able to do in the areas that are important to you?
- Write yourself an honest report about your competencies in the above areas. You can even spend $100–$250 to have yourself tested. The rule of training is to know where you are at!
- The lowest cost method of training is through books (including those available free from the library!). Other low-cost methods are YouTube videos, CDs, and DVDs (also from the library). There are free classes online. Some other courses are also inexpensive, given the benefits they provide.
- Schedule periodic times for your training (like once a month).
- Get out of the office. Go to maintenance trade shows, attend conferences, join associations, and even visit other maintenance facilities when you are on vacation with the family!
- Evaluate your progress. If you don't know the score, you never know if you won the game. Keep evaluating your program, and don't be afraid to change direction.
- This process should never stop. Keep up the learning, and your prospects will be better for it!

Appendix

BIBLIOGRAPHY

Maintenance books by Joel Levitt, all available from Amazon.com.
10 Minutes a Week to Great Meetings, Joel Levitt, Reliabilityweb, Ft Myers, 2017.
10 Minutes a Week to Great Time Management, Joel Levitt, Reliabilityweb, Ft Myers, 2019.
Basics of Fleet Maintenance, Joel Levitt, Reliabilityweb, Ft Myers, 2010.
Conversations in Maintenance, Joel Levitt, Reliabilityweb, Ft Myers, 2019.
Facilities Management, Joel Levitt, Momentum Press, 2013.
Handbook of Maintenance Management, Joel Levitt, Industrial Press, New York, 2009.
Internet Guide for Maintenance Management, Joel Levitt, Industrial Press, New York, 1998.
Lean Maintenance, Joel Levitt, Industrial Press, New York, 2008.
Maintenance Planning, Scheduling, and Coordination, Second Edition, Don Nyman and Joel Levitt, Industrial Press, New York, 2010.
Managing Factory Maintenance, Second Edition, Joel Levitt, Industrial Press, New York, 2004.
Managing Maintenance Shutdowns and Outages, Joel Levitt, Industrial Press, New York, 2004
Quest for Defect Elimination, Joel Levitt, Springfield Resources, PA, 2020.
The Complete Handbook of Preventive and Predictive Maintenance.
TPM Reloaded.

OTHER BOOKS

Coping with Difficult People by Dr. Robert Bramson, published by Anchor Press/Doubleday, New York (HF5548.8.8.B683). This is an excellent book if you commonly face difficult people in your workplace.

Dealing with Difficult People in the Workplace, Employee Productivity Consultants (no longer in business).

Leader Effectiveness Training by Dr. Thomas Gordon, published by Bantam Books, New York. This extends the effective Parents Effectiveness Training to the marketplace. It is an excellent all-around book on the "new" approach to supervision and leadership.

Motivation and Personality, Abraham Maslow, psychologist, Sublime Books (June 10, 2015). He considered this a continuation of his Motivation and Personality, published in 1954.

The One Minute Manager by Drs. Ken Blanchard and Spencer Johnson, published by Berkley Books, New York, n.d. This book takes about 2 hours to read, and it's worth it! These techniques are very useful and surprisingly easy to follow.

Out of the Crisis by W. Edwards Deming, published by MIT Center for Advanced Engineering Study, n.d. This book explains so many things about change, quality, and direction of effort that I unreservedly recommend it. Dr. Deming set the Japanese on a path toward quality that has overwhelmed many of our industries. We should learn the same lessons from one of America's best and most original business systems thinkers.

Planning, Conducting, and Evaluating Workshops, Larry Davis. This excellent text on adult teaching is available from University Associates, Inc. 8517 Production Ave., San Diego, CA 92121. Leading book on adult training and education.

Please Understand Me: Character and Temperament Types, B & D Book Company, 1400 W. 13th Sp. 128, Upland, CA 91786, n.d.

Riding Rockets: The Outrageous Tales of a Space Shuttle Astronaut, Colonel Mike Mullane, NASA Astronaut, Scribner; Reprint edition (February 14, 2006).
The Seven Habits of Successful People, Stephen Covey, Simon and Schuster, 1990.
Shop Class as Soulcraft: An Inquiry into the Value of Work by Matthew B. Crawford, n.d.

MAINTENANCE MANAGEMENT RESOURCES

AFE (Association for Facilities Engineering): includes plant engineering people from the industry, housing, and commercial properties. www.afe.org

ATA (American Trucking Association): 2200 Mill Road, Alexandria, VA 22314. ATA is known for the VMRS (Vehicle Maintenance Reporting Standards). www.truckline.com

New Standard Institute: Providers of computer-based training, seminars, and consultation. 84 Broad, Milford, CT 06460. www.newstandardinstitute.com

Meyers and Briggs (Mother and daughter) are the designers of the personality test referenced. It has been shown to be very useful in dealing with all kinds of people. https://www.myersbriggs.org

RCA RT is a group devoted to the furthering of the thought process on Root Cause Analysis through training and software. www.rcart.com.au

www.reliabilityweb.com is the nexus for a variety of maintenance information and conferences.

SIRF is a membership organization in Australia that tackles issues of its members by running Round Table meetings where peers advise and teach each other and training workshops where they invite in global experts in specific maintenance areas. www.sirfrt.com.au

Springfield Resources. Courses include Maintenance Manager, Computerization of Maintenance, Fleet Maintenance and Buildings and Grounds. All classes are available on-line, in-house with or without consulting support. www.maintenancetraining.com

Work-Sampling is a web site that has training in work sampling and an iPhone App to automate work sampling. www.work-sample.com

SHOP SAFETY AUDIT AND INSPECTIONS

The following are examples of three different general shop safety inspection forms. The Shop Safety Audit list is not exhaustive. The best checklist for your workplace is one that has you developed for your specific needs (and hazards). These are samples to get you started and use where applicable. Whatever the format of your checklist, be sure to provide space for the inspectors' signatures and the date.

It is also useful to have different people go through and act as auditors, including managers, supervisors, and mechanics.

The final checklist is an example of a corrective action plan. It serves as a follow-up to the audits and inspections.

Initial Audit Checklist

Conduct this first audit to make sure all the systems are in place. The review should be done by senior-level personnel, so the systems that are missing are known to management for remediation. The audit should be repeated every few years. It covers the systems of safety that should be already in place.

#	Y	N	NA	First Audit
				Administration
				Are all administrative documents, (e.g., shop safety plan, standard operating procedures (SOPs), SDS) up-to-date and available for review?
				Are emergency phone numbers and the evacuation map posted? Are current after-hours phone numbers posted?
				Does the safety bulletin board contain up-to-date information?
				Is a near miss/close call reporting system in place?
				Are appropriate fire department permits current and posted if required?
				Has a perimeter group been identified (i.e., organizations that share your boundaries), and are there periodic meetings to discuss hazards, coordinated response, etc.?
				Are there appropriate licenses for any activity going on in the site (e.g., asbestos, lead, etc.)?
				Procedures and Training
				Are periodic safety inspections scheduled and done?
				Does initial training include a thorough review of hazards and incidents associated with the job?
				Do all employees know how to get first aid assistance when needed?
				Do the first aiders know when and to which hospital or clinic an injured person should be taken?
				Is adequate instruction in the reason for and use of personal protective equipment (PPE) provided?
				Have all shop users received all necessary training for their work in the shop?
				Are safety training records complete and available upon request?
				Is shop access limited to trained and authorized personnel?
				Is someone trained in first aid and CPR always available?
				On maintenance jobs: Is job safety analysis or safe work procedures in writing?
				Do workers know the symptoms of heat cramps/heatstroke or frostbite/hypothermia?
				Has there been an assessment of the entire facility for hazard types and is PPE available for all hazards encountered in that facility?
				Are manufacturers' manuals kept and followed for all tools and machinery?
				Are all electrical panels accessible, and are the circuits adequately labeled?
				Are there periodic safety drills for possible scenarios?
				Safety
				Are eyewash/showers accessible within 10 seconds (approximately 50 feet), free of obstructions, and have documented inspections?
				Are there written PPE policies for all employees and contractors?
				Are the confined space procedures available for new hires, and has their training been conducted?

(Continued)

#	Y	N	NA	First Audit
				Are resources available to deal with very hot or very cold conditions (e.g., drinking water availability, lined gloves, insulated boots, etc.)?
				Are regular noise surveys conducted? Are the results available?
				Are emergency supplies (e.g., first aid kit, spill kit) available?
				Are first aid supplies replenished after they are used?
				Is there a fire extinguisher inspection and service contract or procedure in place?
				Is there a hoist inspection contract in place?
				Are SDS or material safety data sheets (MSDS) openly available to all employees?
				Are both gender bathrooms provided, and are they equivalently maintained?

If any item is marked "U" for unacceptable or "N" for no, list the appropriate corrective action on the corrective action plan

Annual Inspection Checklist

This safety checklist is a detailed look at the potential areas of hazard. **Important Note:** A lack of hazards does not mean the shop is completely safe since some hazards are created by behavior. But in general, shops with fewer things to trip over will have fewer slips and trips than one that doesn't. A good practice is to rotate the inspector's role among the leadership of the shop.

#	Y	N	NA	Annual Inspection
				Housekeeping, Equipment, and Layout
				Are fire doors unobstructed or wedged open?
				Are aisles, exits, fire doors, and adjoining hallways free of obstructions?
				Have fire extinguishers been inspected in the past year, and are they secured and easily accessible?
				Is emergency lighting adequate and regularly tested?
				Is the shop neat and orderly? Is a place for everything and everything in its place followed? Is the shop clean with no-slip, trip, or fall hazards?
				Are floors free from protruding nails, splinters, holes, and loose boards?
				Are covers or guardrails in place around open pits, tanks, and ditches?
				Are safe walkways and restricted areas clearly marked around shop machinery and dangerous processes?
				Is there adequate (i.e., appropriate brightness, no glare, and the right color for the task), permanent lighting to complete all jobs in the shop safely?
				Are light bulbs for illumination protected from breakage?
				Are heavy items stored low enough to prevent falling on people and for lifting safely?
				Are top-heavy machines, file cabinets, and shelving secured?
				Are work surfaces and grip surfaces safe when wet?
				Are all gas cylinders secured? Do those not in use have caps in place and appropriately segregated?

#	Y	N	NA	Annual Inspection
				Are all blade, shaft, bit, belt, and pulley guards in place to prevent injuries during machine operation?
				Are all shop tools and equipment, both powered and unpowered, maintained in a good state of repair (e.g., sharp, clean, functional)?
				If you handle drums, is there adequate drum handling gear available?
				PPE (available and in good condition for shop users and visitors)
				Are all shop users wearing long pants, sleeved shirts, and appropriate footwear and are loose clothing, long hair, and jewelry restricted while working in the shop?
				Are safety glasses and goggles available and being used?
				Are hard hats worn as required, and are they available for visitors?
				Is noise protection provided for loud work, and is it used?
				Have noise surveys been conducted recently (within one year)?
				Is hand protection being used/worn as required?
				Are welding helmets, gloves, aprons, and curtains available and in use during welding?
				Is there a respirator or proper positive ventilation available?
				Are supplies on hand for incidental chemical spills?
				Hazardous Materials and Communication
				Do SDS and MSDS cover all items in the garage/shop?
				Are all chemicals and chemical waste containers clearly labeled with their contents and primary hazard(s)?
				Are hazardous materials adequately stored, and are storage cabinets appropriately labeled and secured?
				Is the amount of flammable liquids outside of storage cabinets under 10 gallons?
				Are hazardous liquids stored below eye level?
				Are flammable liquids in FM/UL metal safety cans?
				Are flammable liquids storage containers appropriately labeled?
				Are oily rags placed in covered metal containers?
				Are there any unidentified containers of liquids or anything in use in the shop?
				Tools
				Do any temporary repairs on tools, equipment, or facilities have a work order for permanent restoration in the system?
				Are elevating devices used only within their capacity? Are capacities posted on equipment?
				Are the controls of elevating devices the "dead man" type?
				Are portable jacks inspected according to manufacturer requirements?
				Are safety jacks used while working under vehicles?
				Are ladders safe and inspected as appropriate?
				Do extension and straight ladders extend 3' beyond landing?
				Is a stepladder or commercial step stool used for high access?
				Are stepladders used only in the open position?
				Are portable power tools provided with guarding?
				Are portable circular saws equipped with protective guards?

(*Continued*)

#	Y	N	NA	Annual Inspection
				Are unsafe (e.g., damaged, loose, broken) hand tools prohibited?
				Are impact tools and hammers kept free of splinters/mushrooms?
				Are hoists inspected monthly and documented?
				Are hoists inspected annually by an outside service?
				Do impact air tools have safety clips or retainers on them?
				Electrical Safety Issues
				Is stationary shop equipment capable of being locked out for maintenance? Is LOTO used for appropriate tasks?
				Are ground-fault circuit interrupters (GFCIs) used for all portable electrical hand tools?
				Are extension cords in good condition, appropriately rated, and used correctly? Are there no daisy chains, are all prongs intact and no long-term usage?
				Are extension cords rated for hard or extra-hard usage (e.g., three-wire marked = S, ST, SO, STO, SJ, SJO, SJT, and SJTO)?
				Is strain relief intact for all flexible cords and plug fittings?
				Is grounding and bonding integrity maintained for chemical dispensing?
				Is (UL, CSA, DIN, etc.) certified or listed equipment used?
				Are electrical panels labeled appropriately?
				Are electrical panel knockouts in place (e.g., no holes in the panel that are not covered)?
				Are electrical panel access requirements maintained?
				Are all electrical boxes covered?
				Is the pressure washer grounded per National Electrical Code (NEC) requirements?
				Are double insulated or grounded electric power tools used?
				Fall Protection
				Are wall openings and floor holes covered or guarded?
				Is 100% fall protection in place above 6' in height?
				Are employees trained in operating aerial work platforms?
				Are ladders safe and inspected as appropriate?
				Do extension and straight ladders extend 3' beyond landing?
				Is a stepladder or commercial step stool used for high access?
				Do guard rails exist for platforms and scaffolding?
				Hot Work and Welding Safety
				Are compressed gas cylinders securely stored upright with caps?
				Are hot work permits used for grinding, cutting, and welding?
				Are safety and fire watch personnel provided when needed?
				Do oxygen and acetylene torch units have flashback arrestors?
				Do grinders, both portable and stationary, have guards in place?
				Is the stationary grinding wheel tool rest 1/8 inch or less?
				Is the stationary grinding wheel tongue guard 1/4 inch or less?
				Confined Space Entry
				Is the process being followed?
				Are entry and exit procedures adequate?

#	Y	N	NA	Annual Inspection
				Are emergency and rescue procedures in place (e.g., trained safety watchers)?
				Employee Facilities
				Are the facilities kept clean and sanitary?
				Are both gender bathrooms provided, and are they equivalently maintained?
				Are the facilities in good repair?
				Are cafeteria or eating area facilities provided away from hazardous products?
				Are handwashing facilities available, especially in washrooms and near eating areas?

List Item, the Person Responsible and Expected Completion Date; in Status Column, mark as Open, In Process, or Closed

Action Item	Person(s) Responsible	To Be Done By	Status

CORRECTIVE ACTION PLAN

Signature of lead inspector: ________________
Date: ________________________________

ACTION MASTER LIST (AML)

Three techniques to remember seminars, webinars, and books

1. AML (make notes in the three areas).
2. Write a summary report.
3. Take an action that you learned about in the book, class, webinar.

Management Skills

Reference	Idea or short note

People Skills

Reference	Idea or short note

Technical Skills

Reference	Idea or short note

These thirty items comprise your short, medium, and long-term lists of things you want to do to improve yourself as a supervisor and your workplace. Discuss these items in your group. Feel free to add, delete, or change these items after discussion.

General Ideas

ANOTHER APPROACH TO MAKING TRAINING MORE VALUABLE

How to anchor the material from any training, video, or book:

1. Go through the materials and clean up your notes while the training is still fresh.
2. Write a seminar report and discuss the seminar with others.
3. Do something different in your organization as a result of the training.

Review your action master lists and any other notes you have:

Short term action items

Medium-term to follow up

Long term to study

Immediate Action plan: Without action, nothing happens. Without a plan, the work might be meaningless or the wrong direction.

Choose one small thing to do differently when you get back to work. This conscious choice will help anchor the remainder of the material.

____

____

How do you intend to do this?

____

____

____

INTERNATIONAL TRAVELERS EXAMPLES OF CULTURAL FAUX PAS

Albania,	In the locals shake their head to indicate "yes" and nod to indicate "no."
Argentina	It is rude to ask people what they do for a living. Wait until they offer the information.
In Brazil	the sign that is used in North America to mean "okay" means "you're an a-hole."
Bahrain	Never show signs of impatience, because it is considered an insult. If tea is offered, always accept. It is highly inappropriate for a man to touch a woman in public.
Cambodia	Never touch or pass something over the head of a Cambodian, because the head is considered sacred.

(*Continued*)

China	As in most Asian cultures, avoid waving or pointing chopsticks, putting them vertically in a rice bowl, or tapping them on the bowl. These actions are considered extremely rude. Diners pick up the rice bowl and use chopsticks. Accept and give business cards with *both* hands. Study the card first as it represents the person you are meeting. Never write on it or put it in your wallet or pocket, instead use a small card case. When dining, do not start to eat or drink before your host. Avoid embarrassing topics, such as acknowledging Taiwan's independence, freeing Tibet, and Chinese human rights issues. Show deference if someone appears to be senior to you. Allow the Chinese to leave a meeting first. Do not discuss business at meals. If you are bringing gifts, clocks, storks, cranes, handkerchiefs, and anything white, blue or black are definite no-nos because of their association with death. Don't compliment anyone for speaking good English. Chances are, most decision-makers had extensive international exposure abroad. It may also be taken as a sign you cannot find better things to compliment. Stand up when others enter the room When dining with a group and taking food from a common plate, use the implements provided and not your chopsticks or fork, and choose the items closest to you even if you prefer something on the other side of the plate. As a social courtesy, you should taste all the dishes offered, but do not eat all of your meal, or they will assume you did not receive enough food and are still hungry. In many Asian countries, such as China, pointing with the forefinger in public is considered quite rude. In many cultures, one's family defines an individual, says Terri Morrison, co-author of a business etiquette book series, *Kiss, Bow, or Shake Hands*. "Therefore, making an error in a person's name is quite a personal insult." Take the Chinese practice of placing the surname first. "Calling (Chinese President Hu Jintao) 'President Tao' is appalling, like calling him President George, or Bubba," she says. Do not blow your nose in public in Japan, China, Saudi Arabia, or France.
Dominican Republic	When speaking to someone, failure to maintain good eye contact may be interpreted as losing interest in the conversation.
Egypt	Showing the sole of your foot or crossing your legs when sitting is an insult. Never use the thumbs-up sign, because it is considered an obscene gesture.
France	Always remain calm, polite, and courteous during business meetings. Never appear overly friendly, because this could be construed as suspicious. Never ask personal questions. When a top executive of a French company arrived in California on an evening flight, the U.S. company sent a limo to pick up the client at 7:30 p.m., but no staff member went along. The company thought the Frenchman might want to relax at a hotel after a long flight and planned to pick him up the next morning. But the Frenchman was offended because no top official met him at the airport. "The French eat dinner later than 7:30, and he thought he'd be taken out to dinner.". "It nearly killed the deal." During his first meeting in France with his United Kingdom-based company, he drank red wine, then switched to white. "From the expressions of the group, you would have thought I exposed myself," "I later found out you never go to a white after a red, because you can't enjoy the bouquet of the white after you've drunk a red." *Table manners also get many U.S. business travelers in trouble, etiquette experts say. In the USA and the U.K., it's customary to put hands on the lap when not* using them. But it "demonstrates deplorable manners" in France and other countries where the hands and wrists should remain on top of the table. "You also should not leave the table before the meal is over, even if you need to go to the restroom." Do not blow your nose in public in Japan, China, Saudi Arabia, or France.

Greece	If you need to signal a taxi, holding up five fingers is considered an offensive gesture if the palm faces outward. Face your palm inward with closed fingers.
Germany	Avoid discussing sports as it's considered an uneducated thing to do.
	Bush also gave German Chancellor Angela Merkel a shoulder rub while she spoke to Italian Prime Minister Romano Prodi. Many Europeans were offended because the summit was a formal occasion, and they viewed the actions as demeaning.
Indonesia	Both hands are kept on the table while eating at all times.
India	An American offended an Indian engineer working for him in India. The American beckoned the engineer by snapping his fingers and waving a curled forefinger. To an Indian, such a gesture is condescending and insulting. The engineer, who was also upset by the American's use of four-letter words earlier in the week, resigned. "The curled finger proved to be the last straw, and the Indian told me he felt treated like a dog,"
	The client was left-handed and used his left hand to cut a ribbon at a bank opening. "The left hand is considered inauspicious in India, and there was consternation all around."
	"The ribbon was retied and cut again with the right hand."
	Avoid giving gifts made from leather because many Hindus are vegetarian and consider cows sacred. Keep this in mind when taking Indian clients to restaurants. Don't wink, because it is seen as a sexual gesture.
	Don't assume you can use a person's first name—in many parts of the country; it's considered rude.
	Take off your shoes at people's homes, places of worship, and even in some shops and businesses. Rule of thumb: If you see shoes near the door, assume you should take yours off too.
	Don't eat beef.
	Don't accept or give anything with your left hand.
	India: Don't expect people to disagree. "Indians generally don't like to accept that they don't know/understand something...and agree to all that we say". So make sure to dig deeper to make sure you're understood and that what you're asking is doable.
	On the other hand, don't be offended by debate. "We like to argue and debate every small point in any topic/conversation. "It's especially important that people end the conversation feeling they made a few good points."
	Don't refuse hospitality.
	Don't ignore hierarchy in the workplace.
	Failure is not accepted as a part of trying to do something and learning in the process; therefore, doing something or recommending something out of the norm may often not go well.
	Write down your instructions. Verbal communication is treated as uncertain.
Japan	Never write on a business card or shove the card into your back pocket when you are with the giver. This is considered disrespectful. Hold the card with both hands and read it carefully. It's considered polite to make frequent apologies in general conversation.
	During negotiations, the U.S. company invited the Japanese company to Seattle, but for months, officials of the foreign company had trouble obtaining visas. The impatient U.S. company sent high-level executives to Japan to close the deal. It backfired because the Japanese executives were eager to visit the USA and were turned off by the Americans' lack of patience in building a rapport between the companies.
	Picking your teeth with a toothpick in Japan is acceptable.
	Do not blow your nose in public in Japan, China, Saudi Arabia, or France.
Malaysia	If you receive an invitation from a business associate from Malaysia, always respond in writing.
	Avoid using your left hand because it is considered unclean.
Mexico	If visiting a business associate's home, do not bring up business unless the associate does.
Nepal	Major Hindu temples are usually off-limits to foreigners. Don't enter them or take pictures unless given permission.

(*Continued*)

Philippines	Never refer to a female hosting an event as the "hostess," which translates to a prostitute.
Russia	Drinking vodka is a big part of life, and not drinking is considered offensive. Don't perceive traditional Russian hospitality as an attempt to bribe you. Don't shout at people—it's a sign of weakness. Don't interpret the lack of smiles in a general crowd as an unwelcoming attitude. (In fact, a smile in Russia is much more personal, and you will see a lot of smiles when you get closer to people.) Take your shoes off when entering someone's home. Russia Don't put your feet up on the table in front of someone. (Isn't that a significant blunder in **any** country?)
Saudi Arabia	Do not blow your nose in public in Japan, China, Saudi Arabia, or France. It is highly inappropriate for a man to touch a woman in public.
Singapore	If you plan to give a gift, always give it to the company. A gift to one person is considered a bribe.
Spain	Always request your check when dining out in Spain. It is considered rude for wait staff to bring your bill beforehand.
UK	When giving the peace sign, make sure your palm is facing away otherwise; palm facing towards body = f*#$ off. Don't make the mistake of asking personal questions of a Scottish man. He asked the man about his wife and children during a casual conversation. "I was flatly told it was none of my business." "I then asked him about the weather, and I could not get the guy to stop talking about it."
Thailand	In Phuket, Thailand, says several times he's inadvertently insulted or embarrassed Thais, he started the gatherings by talking about business.
Vietnam	Shake hands only with someone of the same sex who initiates it. Physical contact between men and women in public is frowned upon.

PRELIMINARY SELF-TEST IN TIME MANAGEMENT

1. Are you satisfied with the way you spend your time?	yes	no
2. Are you happy with the number of hours you work?	yes	no
3. Do you enjoy your work?	yes	no
4. Do you feel good about the quality of your work?	yes	no
5. Do you effectively cope with the stress/pressure of your job?	yes	no
6. Do you have enough free time?	yes	no
7. Do you enjoy your family?	yes	no
8. Are you satisfied with the results you achieve?	yes	no
9. Are you happy with your bosses' concern for time management?	yes	no
10. Are you satisfied with your subordinate's concern for time management?	yes	no
11. Does your use of time reflect your goals?	yes	no
12. Are you happy with the quality of time with your subordinates?	yes	no
13. Are you an effective delegator?	yes	no
14. Do you react to changes in a constructive manner?	yes	no
15. Are you in control of your telephone?	yes	no
16. Do you take advantage of time windfalls?	yes	no
17. Are you well organized?	yes	no
18. Do you have a well-written plan of organization for your desk?	yes	no
19. Are you in good physical condition?	yes	no
20. Do you occasionally do nothing?	yes	no

Score 1 point for each yes, 0 points for each no	
Score range	
18–20:	Could (and probably should) teach the chapter
15–17:	Good time manager
12–14:	You're above average in time management
9–11:	Pay close attention to this chapter!
0–8:	Tough row to hoe. Is your head still above water?

(Partially adapted from the National Seminar course, "How to get more done")

DAILY ACTIVITIES EXERCISE

Estimate how much time you spend doing various activities. Fill in the percentage of time you spend in a typical week on the activities using the left column.

Please try to note what you do rather than what you'd like to be doing (or what you'd like others to think you're doing).

Activity	**Actual Percent**	**Best Supervisor**
"Wrench turning" time: (All actual physical work activities)	_____	_____
Giving job assignments	_____	_____
Meetings	_____	_____
Circulating (Moving through your domain)	_____	_____
Dealing with users/customers	_____	_____
Teaching, training	_____	_____
Inspecting work	_____	_____
Reading junk mail or E-mail	_____	_____
Hazardous material activities	_____	_____
Other regulatory activities	_____	_____
Scheduling, planning	_____	_____
Seeing salespeople	_____	_____
Budgeting	_____	_____
Purchasing/parts related acts	_____	_____
All other paperwork	_____	_____
Personal activities	_____	_____
(All non-company activities on company time)	_____	_____
Other ______________	_____	_____
Other ______________	_____	_____
Other ______________	_____	_____
Other ______________	_____	_____
Total	**100%**	**100%**

When you return to work, keep a log of what you actually do for a week (use the next two pages as a guide). Time management experts agree the place to start managing your time is finding out what you actually spend your time doing.

Daily Log

Date__/__/__ Name________________________________

Time	Activity	Notes
7:00		
:15		
:30		
:45		
8:00		
:15		
:30		
:45		
9:00		
:15		
:30		
:45		
10:00		
:15		
:30		
:45		
11:00		
:15		
:30		
:45		
12:00		
:15		
:30		
:45		
1:00		
:15		
:30		
:45		
2:00		
:15		
:30		
:45		
3:00		
:15		
:30		
:45		
4:00		
:15		
:30		
:45		

Time	Activity	Notes
5:00	___________	___________
:15	___________	___________
:30	___________	___________
:45	___________	___________
6:00	___________	___________
:15	___________	___________
:30	___________	___________
:45	___________	___________

(We kept it to only 12 hours– you might find it useful to keep a longer log)

Use of the Log Sheets

Copy the log sheets for as many days as you plan to log. The sheet should be filled out as the day progresses (NOT AT THE END OF THE DAY).

Index

J

K

L

M

N

O

P

Q

R

S

T

U

V

W

Printed in the United States
By Bookmasters